MEASUREMENT OF
IMAGE VELOCITY

THE KLUWER INTERNATIONAL SERIES
IN ENGINEERING AND COMPUTER SCIENCE

ROBOTICS: VISION, MANIPULATION AND SENSORS

Consulting Editor: **Takeo Kanade**

MEASUREMENT OF IMAGE VELOCITY

By

David J. Fleet
Queen's University

Foreword by
Allan D. Jepson

Kluwer Academic Publishers
Boston/London/Dordrecht

Distributors for North America:
Kluwer Academic Publishers
101 Philip Drive
Assinippi Park
Norwell, Massachusetts 02061 USA

Distributors for all other countries:
Kluwer Academic Publishers Group
Distribution Centre
Post Office Box 322
3300 AH Dordrecht, THE NETHERLANDS

Library of Congress Cataloging-in-Publication Data

Fleet, David J.
  Measurement of image velocity / by David J. Fleet ; foreword by
Allan D. Jepson.
    p. cm. — (Kluwer international series in engineering and
computer science)
  Revision of the authors's thesis (Ph.D.—University of Toronto)
  Includes bibliographical references and index.
  ISBN 0-7923-9198-5 (acid free paper)
  1. Computer vision. 2. Image processing. 3. Motion—Measurement.
I. Title. II. Series.
TA1632.F62 1992
621.36'7—dc20   91-46859
                                                      CIP

Printed on acid-free paper.

Printed in the United States of America

Contents

Foreword

The beginnings of modern computational vision research can be traced back to Roberts in 1965. Since this time a recurring lesson is that apparently simple visual capabilities in biological systems present rather formidable challenges for machine systems. For example, imagine a scene generated from a moving car. Despite the fact that the images are changing in time at almost every location, we find it trivial to spatially localize other moving objects such as a running child. Over the years computational vision researchers have proposed several schemes for accomplishing this feat, but no scheme has demonstrated the sensitivity and flexibility of human vision. One general approach to this problem, based on the physiology of biological systems, is first to compute the motion across the image of various image structures, and then to infer the three dimensional motion of objects in the scene from these image motions. The lesson mentioned above recurs again for both of these subproblems. In particular, given image motion, the computation of the three dimensional motion of objects in the scene is nontrivial. Perhaps even more surprising is the fact that the accurate measurement of image motion has also presented serious difficulties. In fact, leading researchers in computational vision recently published the opinion that we simply cannot expect to compute accurate image velocities.

This research monograph, which is based on David Fleet's Ph.D. dissertation at the University of Toronto, addresses this currently controversial issue of accurate image motion measurement. Fleet begins with a discussion of image formation, and explains the difficulties behind this apparently simple problem. Next he presents recent approaches to image motion measurement, and highlights the assumptions about image structure used in each of these methods. The mismatches between some of these assumptions and the generally expected image structure are the basis for the criticism of these methods. The key advance in this book is the introduction of a novel method for the measurement of image motion, along with the subsequent analysis of its performance over a range of expected image deformations. The results of an implementation are also used to quantify the performance of the method in different situations.

The issues addressed in this monograph regarding the design and analysis of image measurement tools extend considerably beyond the particular method of motion measurement developed here. For example, the discussion of the stability of phase crossings to small affine deformations of the image have direct relevance to the stability of zero-crossings in bandpass signals. In fact, the tools developed here have already been incorporated into methods for the accurate measurement of other visual primitives such as spatial orientation and binocular disparity. There is recent evidence that a similar approach is used by biological visual systems for the accurate measurement of binocular disparity; so this material is also relevant to physiologists and psychophysicists studying early visual processes.

There is considerable work yet to be done before we can build, for example, a *driver's aide* to help identify the running child mentioned above. I believe that the tools developed in this monograph should provide a rich source of ideas for future research on such an *artificial vision* capability.

Allan Jepson

Preface

In simple terms, vision is the process of seeing, the inference of world properties from the patterns of light sensed by an observer. At first it seems to be a coherent and effortless process, yet upon investigation, vision appears to be a collection of intricate processes that interact in subtle ways to provide a stable, cohesive model of the visual world. The perception of motion, the dynamic relationship between the camera and objects in the scene, is one of our most important visual faculties. It facilitates, among other things, the inference of our movement through the world, the detection and recognition of other moving creatures, the inference of the depth and the surface structure of visible objects, and hand-eye coordination.

Visual motion perception concerns the inference of world properties from sequences of images. The projection of the 3-d motion of objects onto the image plane is called the *2-d motion field*. It is a purely geometric quantity that relates 3-d velocities and positions of points to 2-d velocities and locations in the image plane. From it, with knowledge of the projection, it is possible to infer the 3-d movement of the camera and the local structure of the visible surfaces. The determination of relative 3-d velocity and surface structure has been an active, yet controversial area of research, producing few practical results. One of the main difficulties has been the sensitivity of these methods to the relatively inaccurate estimates of the 2-d motion field that are obtained from image sequences, referred to as optical flow or image velocity.

The goal in measuring *image velocity* is to extract an approximation to the 2-d motion field from a sequence of images. For this, a *definition* of image velocity in terms of spatiotemporal patterns of image intensity and a suitable measurement technique are required. The difficulty is that image intensity depends on several factors in addition to the camera and scene geometry, such as the sources of illumination and surface reflectance properties. Until recently, it has been common to assume that, to a good approximation, patches of intensity simply translate between consecutive frames of an image sequence. The monograph argues that this assumption is overly restrictive. A robust method for measuring image velocity should allow for higher-order geometric deformations (e.g. dilation and rotation), photometric effects (e.g. shading), and the existence of more than one legitimate velocity in a local neighbourhood (e.g. because of specular reflections, shadows, transparency, or occlusion).

Towards this end, the monograph presents a new approach to the measurement of image velocity that consists of three (conceptual) stages of processing. The first constructs a velocity-scale specific representation of the input based on the complex-valued outputs of a family of velocity-tuned filters. Such preprocessing separates image structure according to scale, orientation and speed in order to isolate manifestations of potentially independent scene properties. For the second stage of processing, component image velocity is defined, for each filter output, as the component of velocity normal to level contours of constant phase. Except for isolated regions in which phase-based measurements are unreliable, this definition is shown to yield a dense, accurate set of velocity measurements. In a third stage, these component velocity measurements are combined. However, this last stage involves a host of complex issues relating to scene interpretation, and is examined only briefly in this monograph.

The use of phase information is justified in terms of its robustness with respect to deviations from image translation that exist commonly in perspective projections of 3-d scenes. The occasional instability of phase information is shown to be related to the existence of phase singularities. These neighbourhoods of instability about the singularities are reliably detected with little extra computational effort, which is viewed as an essential component of the resulting technique. Finally, we report a series of experimental results for which an implementation of the technique is applied to both real and synthetic image sequences.

This monograph is a revised version of my PhD dissertation, the successful completion of which was by no means an individual effort. I am most deeply indebted to my supervisor Allan Jepson. Allan's insights, mixed with encouragement, friendship and patience, have been a major influence in more ways that this work admits. I am also grateful to John Tsotsos for his continued support and his efforts in establishing our vision group, to the other members of my examining committee, Peter Burt, Eugene Fiume, Peter Hallett, Geoff Hinton, and Demetri Terzopoulos for their comments on the dissertation, and to the past and present members of the various vision research groups in Toronto for helping to create an exciting environment in which to work and grow. In particular, thanks to Stuart Anstis, John Barron, Gregory Dudek, Ron Gershon, Michael Jenkin, Michael Langer, Niels Lobo, and Evangelos Milios.

Although most of this work was done in Toronto, large portions of it were completed while I was visiting the Cognitive Systems Group at the University of Hamburg. Bernd Neumann and Michael Mohnhaupt were especially generous in arranging these trips, and in providing valuable feedback in many interesting discussions.

I am also grateful to David Heeger for providing his results; to L. Quam for Image-Calc and the Yosemite sequence; to Richard Szeliski and to the vision groups at the University of Massachusetts, NASA-Ames Research Center, and SRI International, for providing some of the images and image sequences used in the experimental work; to Rick Gurnsey for stimulating discussions that continue to interest me; to Michael Jenkin for his stereo implementation used in Chapter 11; to Keith Langley for our recent discussions about phase information; to Yvonne Burgess for useful editorial comments; and to the Universities of Toronto and Hamburg, NSERC Canada, CIAR, and ITRC for financial support.

Finally, warm thanks to other friends and family. I am especially grateful to my mother, for emphasizing the importance of scholastic achievement and for her encouragement; she also read this monograph several times and suggested many useful editorial changes.

David J. Fleet

Part I

Background

Chapter 1

Introduction

Without doubt, our visual perception of motion is fundamental to our understanding of and our interacting with the world. It is one of the primary methods by which we determine the relative motion between different objects, including our own motion (egomotion) through the world. It is also a rich source of information about the depth and surface structure of objects, since the variations in viewpoint caused by motion convey three-dimensional structure in much the same way as two views do in the context of stereopsis. It is thought to be important for passive aspects of scene interpretation and object recognition, as well as tasks required for active exploration.

With respect to passive scene interpretation, visual motion analysis is usually discussed in the context of *scene reconstruction* – computing local descriptions of visible surfaces. Beyond this, the coherence of motion in an image can be used for *segmentation*, the inference of foreground/background relationships and the shapes of objects. For example, the temporal coherence of random dot patterns induces the perception of shape, just as random dot stereograms lead to the perception of depth and shape without readily available spatial features such as edges [Julesz, 1971]. Visual motion can also be a powerful cue that enables us to classify different forms of movement, such as rigid versus nonrigid motion, and to recognize classes of objects on the basis of their motion, whether they are biological organisms or not.

With respect to active sensing in the context of a moving camera, visual motion perception supplies information about the relative distances and velocities of objects, allowing one to predict and avoid collisions. It facilitates the development of hand-eye coordination, the ability to grasp and manipulate objects, and the stabilization of objects on the retina (object tracking) that allows for detailed scrutiny of surface shape and/or textural properties.

How could a machine derive a plausible description of the world, including its dynamic properties, from a temporal sequence of images? Stated in this way, visual motion analysis is no less complex than the analysis of static images, and it shares many of the same problems that exist in relation to the interpretation of other sensory inputs, such as acoustic or tactile signals. A central theme in research efforts to solve these problems is the discovery of structural regularities in the input signals that reliably reflect important properties of the scene [Lowe, 1985; Witkin and Tenenbaum, 1983; Pentland, 1986]. Similarly, it has been important to develop constraints on, and models of, the world that allow the inferential process to proceed from image primitives to scene properties [Zucker, 1981; Pentland, 1986; Jepson and Richards, 1990]. It is the inferential leverage gained by extracting or measuring a particular image property that justifies the necessary computational resources.

This monograph examines the problem of defining and measuring image regularities that reflect spatiotemporal properties of the scene. Because objects in the world are generally coherent and move (or evolve) in constrained ways, temporal variations in image intensity are also constrained and will exhibit certain regularities. The focus of this monograph concerns the measurement of the apparent motion in the image plane that is often called image velocity or optical flow. Ideally, the measured optical flow should coincide with the motion of surfaces projected onto the image plane. This primitive form of spatiotemporal information is a fundamental precursor to almost all aspects of visual motion perception and dynamic scene analysis. Our central thesis is that the phase information contained in the output of velocity-tuned linear filters can be a reliable image property with which to measure velocity. Although we concentrate on motion analysis, our analysis of the use of phase information is pertinent to several other problems in computational vision, especially the measurement of orientation in static images and binocular disparity for stereopsis.

1.1   Visual Motion Perception

When an object moves relative to a camera, points on the surfaces of the object generate trajectories (particle paths) in space-time. The projection of these paths onto the imaging surface produces 2-d paths, the derivatives of which are *2-d velocities* (depicted in Figure 1.1). The collection of these 2-d velocities is commonly referred to as the *2-d motion field* [Horn, 1986]. The geometric nature of the motion field conveys a wealth of quantitative information to the observer about the surface structure of objects (e.g. local depth and orientation), the relative 3-d velocity between objects and the observer (egomotion), and the relative 3-d motion between different objects in the scene (i.e. between objects whose projected motions are not due solely to the observer's motion) [Koenderink and van Doorn, 1976; Longuet-Higgins

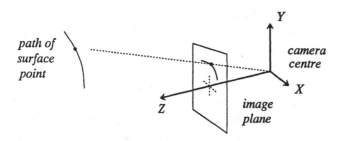

Figure 1.1. 2-D Motion Field: *The 3-d paths of surface points are projected onto the image producing 2-d paths, the instantaneous derivatives of which define the 2-d motion field over the imaging surface.*

and Prazdny, 1980; Waxman and Ullman, 1985].

Computational aspects of *visual motion perception* have been studied extensively as two subproblems, namely, the measurement of 2-d image velocity (also called optical flow), and the extraction of 3-d scene structure and egomotion from the image velocity measurements. By the *measurement of image velocity* we mean the extraction of a local description of the motion field from spatiotemporal patterns of image intensity. The difficulty in measuring image velocity is that image intensity depends on several independent aspects of the image formation process. For example, in the simplified case of a single smooth surface, temporal variations in image intensity may be caused by

- *Perspective Projection:* As a surface moves in 3-d, the relative image locations to which nearby points on a surface project undergo geometric deformations from frame to frame. Translation occurs as surfaces move across the camera's line of sight. Rotation and dilation occur as surfaces rotate, or move toward the camera, etc.;

- *Photometric Effects:* Shading and mirror-like reflection are common properties of surface reflectance, and significantly affect image intensity as a function of time. Variations in illumination conditions can be significant also;

- *Camera Distortions:* The sensors may also cause variations in image intensity, for example, due to spatiotemporal smoothing (motion blur) or gain control.

Because of these different sources of intensity variation, it is unlikely that the image intensity corresponding to a single surface point will remain the same at all times. Therefore, by tracking points of constant intensity, one will not, in general, produce satisfactory estimates of the 2-d motion field.

Indeed, the fundamental problem is to *define* image velocity, and to devise a complementary measurement technique that yields a good approximation to the 2-d motion field. Another way to view the measurement of image velocity is in terms of tracking an image property (or feature) from frame to frame. Our goals in this context are to specify an image property that evolves according to the motion field, and to develop a robust tracking technique.

At present, the most widely held definitions of image velocity are based on a model of temporal intensity variation in terms of local image translation, the assumption being that to a good approximation regions of the image simply translate from frame to frame. The most widely used measurement techniques that follow from this model can be classified as differential techniques [Fennema and Thompson, 1979; Horn and Schunck, 1981; Nagel, 1983; Glazer, 1987], matching techniques [Burt et al., 1983; Anandan, 1989; Little et al., 1988], feature-based approaches [Marr and Ullman, 1981; Buxton and Buxton, 1984; Waxman et al., 1988], and energy-based techniques [Adelson and Bergen, 1986; Heeger, 1988; Shizawa and Mase, 1990]. Such approaches have been criticized on the grounds that the velocity measurements do not accurately reflect the projected motion field in projections of typical 3-d scenes [Verri and Poggio, 1987]. In particular, Verri and Poggio (1987) claimed that accurate estimates of the projected velocity field are generally inaccessible, and that only qualitative information concerning the motion field can be measured. Given the sensitivity of known techniques for inferring scene structure from velocity measurements (e.g. [Barron et al., 1991; Jepson and Heeger, 1990]), this claim is a serious challenge to the widely held view that the study of visual motion would lead to techniques for the computation of quantitative scene properties.

1.2   Outline of Monograph

This monograph addresses the problem of 2-d velocity measurement. The emphasis is on developing a computational technique that yields a reliable, local description of the motion field in terms of instantaneous 2-d translation. The technique produces accurate measurements given image translation, and it is robust with respect to deviations from image translation that commonly occur in image sequences, such as geometric deformation and shading variations. Other important issues include noise robustness, localization in space-time, and the ability to discern different velocities within a single neighbourhood, which is important in cases of specular reflections, shadows, transparency, and near occlusion boundaries.

Our approach consists of three stages of processing:

- decomposition of the input (in space-time) according to a family of velocity-tuned linear filters;

- utilization of the local spatiotemporal *phase* behaviour of the band-pass filter outputs to measure *component image velocity*: the component of 2-d velocity normal to spatially oriented structure in the image;

- integration of these local measurements (e.g. toward an explicit representation of the 2-d velocity field).

The monograph is divided into four main parts, each of which contains several chapters. Part I provides the introduction and an extensive review of the background material for the research that follows. It formulates the problem of velocity measurement in terms of the expected time-varying nature of image intensity, concentrating on contrast changes, geometric deformation and multiple local velocities. In addition, it reviews frequency analysis in space-time, the design of velocity-tuned filters, and the major existing approaches to velocity measurement.

Part II presents a new phase-based method for measuring image velocity, and demonstrates its performance with extensive experimentation using real and synthetic image sequences. Part III then provides a theoretical basis for the success of phase-based techniques. It addresses the *stability* of phase information with respect to deviations from a model of image translation, and a common form of *instability* related to phase singularities. Part IV provides a summary and discussion of the main results.

Parts II and III contain the major contributions of the research, and are relatively self-contained. They could be read directly, referring to earlier chapters for occasional clarification or reference.

1.3 Notational Preliminaries

This section outlines most of the mathematical notation that is used throughout the monograph. It is presented here to avoid repetition.

The image $I(\mathbf{x}, t)$ (or $I(\mathbf{x})$) is taken to be a real-valued function of space-time (or space, respectively). Vectors are denoted using bold-face letters as in $\mathbf{x} \equiv (x, y)^T \equiv (x_1, x_2)^T$, where $\mathbf{x}^T$ denotes the transpose of $\mathbf{x}$, $\mathbf{x}^\perp$ denotes the vector perpendicular to $\mathbf{x}$ with the same magnitude, and $\| \mathbf{x} \|$ denotes the L_2 norm of $\mathbf{x}$, that is, $\| \mathbf{x} \|^2 = \mathbf{x}^T \mathbf{x}$.

Linear Operators [Horn, 1986; Bracewell, 1978]

Shift-invariant linear operators are used often in what follows. They are characterized in terms of an impulse response function and convolution, or in the Fourier domain in terms of a transfer function. The convolution of two functions is given by

$$R(\mathbf{x}) = [K * f](\mathbf{x}) \equiv \int_{-\infty}^{\infty} K(\mathbf{x} - \mathbf{w}) f(\mathbf{w}) \, d\mathbf{w} , \qquad (1.1)$$

where $*$ denotes the convolution operator. The filter's impulse response function (also referred to as the filter kernel or the support function) is defined as the filter's output in response to a Dirac delta function $\delta(\mathbf{x})$ which satisfies

$$\delta(\mathbf{x}) = 0 , \quad \text{for} \quad \| \mathbf{x} \| \neq 0 , \qquad (1.2a)$$

$$\int_{-\infty}^{\infty} \delta(\mathbf{x} - \mathbf{x_0}) K(\mathbf{x}) \, dx = K(\mathbf{x_0}) , \qquad (1.2b)$$

for sufficiently smooth functions $K(\mathbf{x})$. The filter kernel and its output response will often be denoted by $K(\mathbf{x})$ and $R(\mathbf{x})$ respectively.

The partial derivative of $I(x, y)$ with respect to x is denoted by $I_x(\mathbf{x}) \equiv \partial I(\mathbf{x}) / \partial x$. The gradient and Laplacian operators are given by

$$\nabla I(\mathbf{x}) \equiv (I_x(\mathbf{x}), I_y(\mathbf{x}))^T ,$$

$$\nabla^2 I(x, y) \equiv \frac{\partial^2 I(x, y)}{\partial x^2} + \frac{\partial^2 I(x, y)}{\partial y^2} .$$

It will also be common to specify derivatives with respect to several dimensions using $I_{\mathbf{x}}(\mathbf{x}, t) \equiv (I_x(\mathbf{x}, t), I_y(\mathbf{x}, t))^T$. The notation $\nabla I(\mathbf{x})|_{\mathbf{x}=\mathbf{x_0}}$ denotes the derivatives of $I(\mathbf{x})$ evaluated at $\mathbf{x_0}$.

Fourier Transform [Bracewell, 1978; Papoulis, 1968]

The Fourier transform of a real or complex function on $\mathbf{x}$, $\mathcal{F}[I(\mathbf{x})] \equiv \hat{I}(\mathbf{k})$, and its inverse $\mathcal{F}^{-1}[\hat{I}(\mathbf{k})] = I(\mathbf{x})$ are given by

$$\hat{I}(\mathbf{k}) = \int_{-\infty}^{+\infty} I(\mathbf{x}) \, e^{-i \mathbf{k}^T \mathbf{x}} \, dx , \qquad (1.3a)$$

$$I(\mathbf{x}) = \frac{1}{(2\pi)^d} \int_{-\infty}^{+\infty} \hat{I}(\mathbf{k}) \, e^{i \mathbf{k}^T \mathbf{x}} \, dk , \qquad (1.3b)$$

where $\mathbf{x} \in \mathbb{R}^d$. The variables $\mathbf{k}$ and ω are often used to denote spatial frequency and temporal frequency. All frequency variables are written in *radians* per unit distance (e.g. seconds). To convert to *cycles* per unit distance, simply divide by 2π.

The Fourier transforms of real signals, as well as the output of linear filters, are generally complex-valued. Let the real and imaginary parts of a complex number z be denoted by $\text{Re}[z]$ and $\text{Im}[z]$, and let $z^* = \text{Re}[z] - i\text{Im}[z]$ denote the complex conjugate of z. We can also express z in polar coordinates as in $z = \rho e^{i\phi}$, where the amplitude and principal phase angle (in radians) are denoted by ρ and ϕ and are given by (e.g. [Priestley, 1985])

$$\rho \equiv |z| \equiv \sqrt{(\text{Re}[z])^2 + (\text{Im}[z])^2} \, , \tag{1.4a}$$

$$\phi \equiv \arg(z) \equiv \text{Im}[\log(z)] \, . \tag{1.4b}$$

Given an arbitrary phase ψ, we denote its principal part by $[\psi]_{2\pi}$ such that $[\psi]_{2\pi} \in (-\pi, \pi]$.

The following properties of the Fourier transform are used occasionally:

- Parseval's theorem:

$$2\pi \; <f(\mathbf{x}), g(\mathbf{x})> \;\; = \;\; <\hat{f}(\mathbf{k}), \hat{g}(\mathbf{k})> \, , \tag{1.5}$$

where the inner product $<\cdot, \cdot>$ is defined by

$$<f(\mathbf{x}), g(\mathbf{x})> \;\; = \;\; \int_{-\infty}^{\infty} f(\mathbf{x})^* g(\mathbf{x}) \, d\mathbf{x} \, . \tag{1.6}$$

Accordingly, $\| f(\mathbf{x}) \|^2 \; = \; <f(\mathbf{x}), f(\mathbf{x})>$.

- convolution theorem:

$$\mathcal{F}[f * g] \;\; = \;\; \hat{f}(\mathbf{k})\,\hat{g}(\mathbf{k}) \tag{1.7}$$

- shifting property:

$$\mathcal{F}[f(\mathbf{x} - \mathbf{x}_0)] \;\; = \;\; e^{-i\mathbf{k}^T\mathbf{x}_0} \, \hat{f}(\mathbf{k}) \tag{1.8}$$

In particular, note that $\mathcal{F}[\delta(\mathbf{x} - \mathbf{x}_0)] \; = \; e^{-i\mathbf{k}^T\mathbf{x}_0}$.

- modulation property:

$$\mathcal{F}\left[e^{i\mathbf{k}_0^T\mathbf{x}} f(\mathbf{x})\right] \;\; = \;\; \hat{f}(\mathbf{k} - \mathbf{k}_0) \tag{1.9}$$

- differentiation:

$$\mathcal{F}\left[\frac{\partial^n f(\mathbf{x})}{\partial x_j{}^n}\right] \;\; = \;\; (i\,k_j)^n \, \hat{f}(\mathbf{k}) \tag{1.10}$$

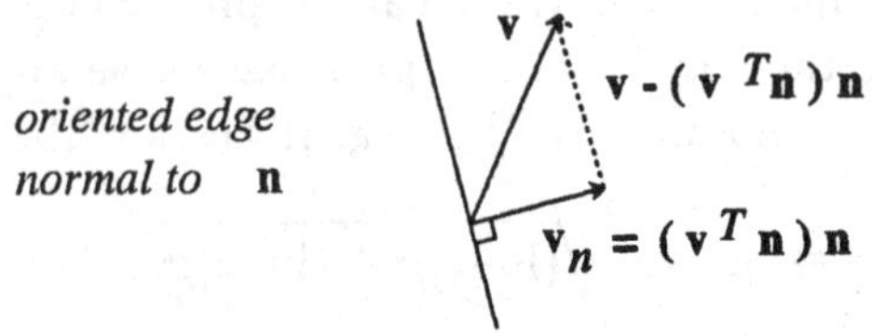

Figure 1.2. Component Velocity: *The component of the 2-d velocity* v *in the direction* n *perpendicular to an oriented edge is referred to as a normal or component velocity* v_n. *The tangential velocity is* $v - (v^T n)n$.

2-D Image Velocity

Throughout the dissertation v is used to denote 2-d image velocity in pixels/frame with speed $v = \| v \|$; v is generally a function of image location and time. Similarly, as shown in Figure 1.2, v_n denotes the component of v in the direction n; that is, $v_n = (v^T n) n = v_n n$ where n is a unit vector. Finally, $\tilde{v}$ denotes an estimate of v.

Velocity refers to change in space as a function of time. For convenience, velocity will sometimes be expressed as a unit direction vector in space-time (i.e. a unit vector tangent to a space-time particle path). The unit direction vector that corresponds to an arbitrary 2-d velocity v is given by

$$ u = \frac{(v_1,\ v_2,\ 1)^T}{\sqrt{v_1^2 + v_2^2 + 1}} . \tag{1.11} $$

It will also be convenient to express u in spherical coordinates (θ_x, θ_v)

$$ u = (\cos\theta_x \sin\theta_v,\ \sin\theta_x \sin\theta_v,\ \cos\theta_v)^T , \tag{1.12} $$

where $\theta_x = \arctan(v_2/v_1)$ gives the spatial direction of motion, and $\theta_v = \arctan(v)$. The expression of velocity in angular coordinates naturally complements the notion of velocity as space-time orientation. Moreover, the angle between two velocities u and ũ is a convenient measure of the distance (i.e. error) between them.

Chapter 2

Time-Varying Image Formation

Vision is the process of inferring scene/world properties from images. Our specific concern is with image sequences and spatiotemporal variations in image intensity. It is therefore natural to begin with a discussion of image formation, the ways in which properties of the world are manifested in images as a function of time and space. In other words, what do we expect the time-varying signal to look like? By examining this question we can begin to identify some of the major issues concerning the measurement of image velocity.

This chapter begins with a definition of the quantity to be measured, the *2-d motion field*. It then discusses the forms of spatiotemporal intensity structure that result from a relatively simple scene model incorporating smooth surfaces with both diffuse and specular components of reflection. The model serves to emphasize the common occurrences of local geometric deformation, smooth contrast variations, and multiple image velocities in a local image neighbourhood. This is an important departure from other approaches, most of which are based on a model of image translation and the assumption of smooth and unique 2-d velocity fields.

2.1   2-D Motion Field

As discussed above, a primary goal of early motion analysis is to detect and measure the variations in the projected image locations of physical surface points as the camera and/or objects in the scene move. When an object moves with respect to the camera, a point on the surface of the object can be thought to trace out a 3-d particle path as a function of time (see Figure 1.1). Its derivative yields the instantaneous 3-d velocity of the point. The projection of the point and its path onto the image plane yields a 2-d

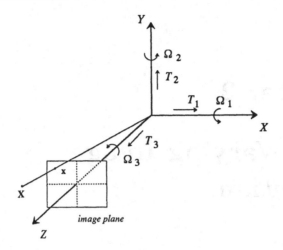

Figure 2.1. Imaging Geometry: *The X, Y, and Z scene coordinate axes, the image plane, and the instantaneous translation (T_1, T_2, T_3) and rotation parameters $(\Omega_1, \Omega_2, \Omega_3)$ are shown. Also depicted is the location of a 3-d scene point* **X** *and the image location* **x** *to which it projects.*

particle path, the derivative of which is the instantaneous 2-d velocity. Following Horn (1986), the *2-d motion field* is defined to be the collection of 2-d velocities on the imaging surface as a function of space and time.

Following Longuet-Higgins and Prazdny (1980), we assume a Euclidean, camera-centred coordinate system for the world with the focal point of the camera at the origin. Scene locations are expressed as vectors in $\mathbb{R}^3$, and are denoted by capital letters, for example $\mathbf{X} = (X, Y, Z)$. Furthermore, as shown in Figure 2.1, we let the optical axis (the line of sight) coincide with the Z-axis, so that the imaging surface lies in the plane $Z = f$ normal to the line of sight, where f is the focal length. Assuming an ideal pin-hole camera model, and therefore perspective projection, the image location $\mathbf{x}$ that corresponds to an arbitrary scene point $\mathbf{X}$ is given by

$$(x, y, f)^T \;=\; \frac{f}{Z}(X, Y, Z)^T \,. \tag{2.1}$$

We assume for notational convenience that $f = 1$, and the third component of $(x, y, f)^T$ can be dropped so that image locations are written as vectors in $\mathbb{R}^2$.

The relative motion between a camera and a set of rigidly moving surface points can be expressed in terms of instantaneous translation and rotation of the camera [Longuet-Higgins and Prazdny, 1980]. As shown in Figure 2.1, let $\mathbf{T} = (T_1, T_2, T_3)^T$ denote the components of translation along the three coordinate axes, and let $\mathbf{\Omega} = (\Omega_1, \Omega_2, \Omega_3)^T$ denote the components of camera rotation about each axis. The angle of rotation per frame, assumed to be small, is given by $\| \mathbf{\Omega} \|$ (in radians), and the axis of rotation is $\mathbf{\Omega}/\| \mathbf{\Omega} \|$. The velocity of a scene point relative to the camera can be expressed as

$$\mathbf{V} = -\mathbf{T} - \mathbf{\Omega} \times \mathbf{X}. \tag{2.2}$$

Differentiating (2.1), with $\mathbf{V} \equiv d\mathbf{X}/dt$ given in (2.2), we obtain the standard expression for the instantaneous projected velocity of a surface point $\mathbf{X}$ at location $\mathbf{x}$ in the image plane:

$$\mathbf{v} = Z^{-1} \begin{pmatrix} -T_1 + x\,T_3 \\ -T_2 + y\,T_3 \end{pmatrix} + \begin{pmatrix} \Omega_1 xy - \Omega_2(1+x^2) + \Omega_3 y \\ \Omega_1(1+y^2) - \Omega_2 xy - \Omega_3 x \end{pmatrix} \tag{2.3}$$

For specific surface models, such as planar or quadratic surface patches, Z^{-1} can be replaced by an appropriate function in terms of surface parameters and image coordinates (e.g. [Waxman and Wohn, 1985]).

In the case of a smooth (differentiable) surface and constant $\mathbf{T}$ and $\mathbf{\Omega}$, the motion field in the image neighbourhood of $\mathbf{x}_0$ can be expressed, to first-order, as

$$\mathbf{v}(\mathbf{x}) = \mathbf{v}(\mathbf{x}_0) + A\,(\mathbf{x} - \mathbf{x}_0) + O(\| (\mathbf{x} - \mathbf{x}_0) \|^2), \tag{2.4}$$

where $A = \nabla \mathbf{v}(\mathbf{x})|_{\mathbf{x}=\mathbf{x}_0} \in \mathbb{R}^{2\times2}$. The four elements of A, that is a_{ij}, depend on the components of 3-d translation and rotation, T_n and Ω_m, as well as the surface structure of the object (e.g. orientation and curvature). Intuitively, they combine to specify the amount of rotation, dilation and shear as the surface rotates or moves toward (or away from) the camera [Koenderink and van Doorn, 1976; Longuet-Higgins and Prazdny, 1980]. For example, the divergence is given by $a_{11} + a_{22}$. In general, there are four degrees of freedom in A; its elements are independent and typically non-zero. Even with planar surfaces the motion field is quadratic in $\mathbf{x}$, and, except in special cases, $a_{ij} \neq 0$ [Waxman and Ullman, 1985; Adiv, 1985]. In other words, pure image translation rarely occurs, and we can expect some variation in $\mathbf{v}$ within local image neighbourhoods.

2.2 Time-Varying Image Intensity

The motion field is a purely geometric quantity relating the locations and velocities of points on objects to image locations and image velocities. But it must be measured or approximated from spatiotemporal variations in image intensity. Because image intensity is a function of several factors in addition to the scene and camera geometry, formal definitions of *image velocity*, often called optical flow, have been elusive.

There are two principal factors that influence image intensity and its variation through space-time: 1) photometry; and 2) the imaging and scene geometries[1] [Horn, 1986; Horn and Sjoberg, 1979; Verri and Poggio, 1987; Nagel, 1989]. In what follows, let p refer to a location on a smooth surface in a local, surface-based coordinate system. Then, let $x(p, t)$ denote the image location corresponding to the perspective projection of p at time t, where $x_0(t) = x(0, t)$ denotes the image coordinates of the projection of $p = 0$, the origin of the local surface coordinate system. Conversely, let $p(x, t)$ be the surface location at time t that projects to the image at x.

2.2.1   Geometric Deformation

The variations of image intensity that are due to imaging and scene geometry depend on the instantaneous location and velocity of the surface relative to the camera. For a smooth surface, we can express the change in the image locations to which fixed points on the surface project as a smooth deformation, that is,

$$x(p, t) \;=\; d(x(p, 0), t) \,. \tag{2.5}$$

For example, pure translation and isotropic dilation are given by

$$d(x(p, 0), t) \;=\; x(p, 0) \,+\, v\,t \,, \tag{2.6a}$$

$$d(x(p, 0), t) \;=\; x(p, 0) \,+\, \alpha\,t\,(x(p, 0) - x_0(0)) \,, \tag{2.6b}$$

where $v\,t$ is the translation, and αt determines the amount of dilation. In the general case, following (2.4), we can express the projected image locations of surface points to first-order in space-time, about $p = 0$ and time $t = 0$, as follows:

$$\begin{aligned}
x(p, t) \;=\;& x(p, 0) \,+\, t\,[\,v(x_0(0), 0) \,+\, A\,(x(p,0) - x_0(0))\,] \\
& + O(t^2 + t\Delta x^2) \,,
\end{aligned} \tag{2.7}$$

where $A = v_x(x, t)\,|_{x=x_0, t=0} \in \mathbb{R}^{2\times 2}$, and $\Delta x = \|\, x(p, 0) - x_0(0)\,\|$. Equation (2.7) implies that local image patches can be expected to translate (by $v(x_0(0), 0)$) and deform (according to A). As discussed in Section 2.1, (2.7) amounts to a general first-order deformation. For reasonably smooth surfaces and general viewing conditions, for which the surface is a reasonable distance from the camera, the amount of image deformation caused by dilation, rotation and shear (i.e. $\|\, A\, \|$) is typically small compared to $\|\, v(x_0)\, \|$.

---

[1]Other issues such as spatiotemporal sensor integration and sampling rates are discussed in Chapter 3.

2.2.2 Photometric Factors

The photometric factors that influence image intensity are difficult to model accurately: they include the principal sources of illumination, the reflectance properties of different surfaces, and the scene geometry in order to account for mutual illumination and shadows. Therefore, rather than attempt to provide a detailed physical model for the general case, we sketch a model in which all but the simplest aspects of illumination and reflectance are ignored. The model consists of a smooth surface in 3-d that moves with respect to the camera. The surface reflectance function is assumed to be a linear combination of a perfectly diffuse (Lambertian) term and a specular term. Although simple, this model demonstrates the influence of shading and specular reflection on image intensity as a function of image location and time. A more realistic model may complicate, but will not negate these effects.

In the following sections we summarize the results that are of immediate interest; the details can be found in Appendix A. Because the diffuse and specular terms are assumed to combine linearly, their time-varying behaviour is considered separately.

Diffuse Component

For a perfectly diffuse (Lambertian) reflecting surface, the scene radiance at a given time depends only on the angles between the surface normal and the sources of illumination, and not on the orientation of the camera. As the object moves or as the sources of illumination change through time, the shading of the surface also changes. More precisely, it is shown in Appendix A that the spatiotemporal variations in intensity caused by the motion of a smooth Lambertian surface can be modelled as

$$I(\mathbf{x}(\mathbf{p}, t)) \;=\; S(\mathbf{p}, t)\, I(\mathbf{x}(\mathbf{p}, 0), 0) \;, \tag{2.8}$$

where $S(\mathbf{p}, t)$ is a shading function defined on the surface, and $\mathbf{x}(\mathbf{p}, t)$ is given by (2.7). For reasonably general lighting conditions, the shading function will also be smooth (in space and time), and can be expressed as a Taylor series. For example, we can express S to first-order, about $\mathbf{p} = 0$ and $t = 0$ for convenience, as follows:

$$S(\mathbf{p}, t) \;=\; 1 \;+\; t\, S_t|_{\mathbf{p}=0, t=0} \;+\; \mathbf{p}^T S_{\mathbf{p}}|_{\mathbf{p}=0, t=0} \;+\; O(\|\, (\mathbf{p}, t)\, \|^2)\;. \tag{2.9}$$

The second term in (2.9), $t\, S_t$, represents a (spatially) homogeneous scaling of intensity (contrast variation). It is due primarily to a change in angle between the principal light sources and the surface normal. The next term, $\mathbf{p}^T S_{\mathbf{p}}$, represents the shading gradient. It depends primarily on the curvature of the surface and on the variation in surface irradiance due to nearby light sources. In the case of nearby sources, the local surface irradiance depends on the distance between the surface point and the light source, which causes a shading gradient even for planar surfaces.

Specular Component

Variations in image intensity are also caused by *specular* (direction-specific) components of reflectance. Unlike diffuse reflectance, the surface radiance associated with specular reflection depends on the angle between the surface normal and the viewing direction. A perfect specularly reflecting surface acts like a mirror, reflecting the light from each incident direction to a single emittant direction, determined by the illumination direction and the surface normal. In effect, specular reflections create *virtual* surfaces, which may be indistinguishable from real surfaces. As discussed in Appendix A, when there is motion of the camera, the reflecting surface, or the reflected surface, these virtual surfaces give rise to *virtual 2-d motion fields*. Moreover, these virtual motion fields will be different from the 2-d motion field that corresponds to the reflecting surface.

In practice, most surfaces are not perfect specular reflectors, for which incident light from a specific direction is reflected to only one direction. More commonly, the specular component of reflection emits light over a range of directions that lie reasonably close to the direction of perfect mirror reflection [Phong, 1975; Horn, 1977]. In terms of the microfacet model described in [Cook and Torrance, 1982], it is uncommon to encounter surfaces for which all the specularly reflecting microfacets of the surface are aligned (or parallel) in surface neighbourhoods. The rougher the surface, or the broader the distribution of mircofacet orientations about the surface normal, the greater the blurring of the reflected pattern [Horn, 1977].

Several other consequences of specular reflection are worth mentioning. First, when the reflected surface is distant (far from the reflecting surface) the specular component of surface radiance (reflection) is extremely sensitive to changes in the viewing direction and surface orientation. Therefore, any motion of the reflecting surface can induce extremely fast 2-d velocities in the virtual motion field. This also means that even slight variations of the surface normal over a smooth surface can cause significant non-rigid distortion of the projected pattern. Second, note that because of the significant power of the specular reflection in comparison to the diffuse component, and its sensitivity to changes in surface reflectance properties and perturbations of the surface normal, specular reflection often tends to enhance the visibility of certain properties of the actual (physical) reflecting surface. This can be seen, for example, in the highlights off a polished tile floor, where cracks between the tiles involve variations in the surface normal and a change in reflectance properties if, for example, they are filled with dirt.

2.3   Discussion

The purpose of velocity measurement is the recovery of a description of the motion field in terms of local instantaneous 2-d velocity (translation). The simple model of intensity variation outlined in this chapter helps to raise several important issues concerning definitions of image velocity, and suitable techniques for its measurement.[2]

First, for diffusely reflecting surfaces, we expect local variations in intensity due to general geometric deformation, as well as smooth contrast variations due to photometric (shading) influences. Although the factors taken into account in deriving this model are minimal, they are sufficient to illustrate the primary sources of intensity variation, namely, geometric deformation and smooth contrast variations. Other factors, required by a more realistic model, would introduce further sources of distortion. For example, contrast variations are also caused by sensor gain control, spatiotemporal integration (smoothing), mutual illumination, and shadows. In any case, in order to produce reliable estimates of image velocity, the measurement technique must *1)* give accurate estimates of image translation, and *2)* be robust in the face of first-order geometric deformation and smooth contrast variation.

Secondly, although the motion fields with which we are primarily concerned correspond to the motion of physical surfaces, it is difficult, without scene interpretation or *a priori* constraints, to distinguish between the time-varying intensity behaviour of real surfaces and virtual surfaces. Moreover, the local variation of image intensity often reflects the motion of more than one surface (motion field). This is true where intensity is a function of the reflectance properties of both the physical surface and a reflected (virtual) surface. Other situations in which more than one legitimate 2-d velocity can exist in a local neighbourhood include transparency (transmission of light through material), shadows under diffuse or ambient illumination conditions (cloudy or foggy days) for which the envelope of the shadow can be seen moving over the actual surface, and occlusion. The existence of multiple image velocities in a single image neighbourhood stands in stark contrast to the conventional assumption of a unique and smooth 2-d velocity field.

---

[2]The subsequent process of inferring scene structure, *viz.* the physical causes of spatiotemporal intensity variations, are important, but are beyond the scope of the monograph.

Chapter 3

Image Velocity and Frequency Analysis

This chapter describes the spatiotemporal input signal and models of image velocity using Fourier analysis. It is shown that velocity is a form of orientation in space-time, and has a very simple expression in the frequency domain. We begin with the 2-d translation of textured patterns and 1-d profiles, and then discuss the effects of spatiotemporal localization (windowing), the uncertainty relation, transparency, occlusion, temporal smoothing, sampling and motion blur. This perspective is also important for the design and understanding of linear filters which are characterized by the regions in the frequency domain to which they respond strongly or attenuate. Linear filters can be used as an initial stage of processing to separate image structure according to scale and velocity. The construction of a representation of the input based on a *family* of *velocity-tuned* filters is discussed in Chapter 4.

3.1 2-D Image Translation

2-D translation is perhaps the simplest form of intensity variation that is commonly used to model the local temporal variations in an image sequence. Let $I(\mathbf{x}, t)$ be composed of a 2-d intensity function $I_0(\mathbf{x})$ translating with velocity $\mathbf{v}$; that is,

$$I(\mathbf{x}, t) \;=\; I_0(\mathbf{x} - \mathbf{v}t) . \tag{3.1}$$

The Fourier transform of (3.1), written $\hat{I}(\mathbf{k}, \omega)$, can be derived using the Fourier shift property (1.8) as follows:

$$\begin{aligned}
\hat{I}(\mathbf{k}, \omega) &= \int\int I_0(\mathbf{x} - \mathbf{v}t)\, e^{-i(\mathbf{x}^T\mathbf{k} + t\omega)} dx\, dt \\
&= \hat{I}_0(\mathbf{k}) \int e^{-it\mathbf{v}^T\mathbf{k}}\, e^{-it\omega}\, dt
\end{aligned}$$

$$= \hat{I}_0(\mathbf{k}) \, \delta(\omega + \mathbf{v}^T \mathbf{k}) \,, \qquad\qquad (3.2)$$

where $\hat{I}_0(\mathbf{k})$ is the Fourier transform of the 2-d intensity pattern $I_0(\mathbf{x})$.

Equation (3.2) shows that when the translating pattern is expressed in the frequency domain, all of its non-zero power (its non-zero Fourier coefficients) lies on a plane containing the origin; that is, $\delta(\omega + \mathbf{v}^T \mathbf{k})$ is non-zero only when $\omega = -\mathbf{v}^T \mathbf{k}$. The speed $\| \mathbf{v} \|$ determines the angle between the two planes $\omega = -\mathbf{v}^T \mathbf{k}$ and $\omega = 0$. The direction of $\mathbf{v}$ determines the orientation of the velocity plane about the ω-axis. Finally, there is a one-to-one correspondence between finite 2-d image velocities $\mathbf{v}$ and those planes that intersect the origin in frequency space but which do not contain the entire ω-axis.

3.2   2-D Component Velocity

The usefulness of 2-d image translation as a model for image velocity is somewhat limited because most techniques aim to measure velocity only in relatively narrow spatiotemporal apertures. With narrow apertures, smoothly varying velocity fields can be estimated based on translational image velocity as opposed to more complex descriptions of the motion field over larger neighbourhoods. Similarly, localization helps ensure good resolution of the velocity field. That is, it is unnecessary to try to determine where within the window of measurement support the velocity estimate is relevant. Localization also helps to minimize (isolate) the adverse effects of occlusion boundaries on velocity measurement, and it helps to ensure that the measurements of spatially disjoint objects will remain independent.

However, if measurements are restricted to narrow apertures, the intensity structure upon which the measurements are based will often be one-dimensional (edges). In this case we can only expect to measure the normal component of the 2-d velocity reliably, that is, the component of $\mathbf{v}$ in the direction $\mathbf{n} = (\sin\theta, -\cos\theta)$, given by $\mathbf{v}_n = v_n \mathbf{n} = (\mathbf{v}^T \mathbf{n})\,\mathbf{n}$, where $\mathbf{n}$ is perpendicular to the local, oriented intensity structure with orientation θ (see Figure 1.2). This is referred to as *the aperture problem* [Marr and Ullman, 1981].

Consider a 1-d intensity profile $I_0(x)$ with orientation θ, translating with component velocity $\mathbf{v}_n = v_n \mathbf{n}$; that is,

$$I(\mathbf{x}, t) = I_0(\mathbf{x}^T \mathbf{n} - t v_n) \,. \qquad\qquad (3.3)$$

As with (3.2), the Fourier transform of (3.3) can be derived using the Fourier shift property. Toward this end, let $\mathbf{y}$ be defined by the orthogonal rotation $\mathbf{y} = N\mathbf{x}$ where N and N^{-1} are defined with $\mathbf{n}$ and $\mathbf{n}^{\perp}$ as row and column vectors; that is,

$$N = \begin{bmatrix} \mathbf{n}^T \\ (\mathbf{n}^\perp)^T \end{bmatrix}, \qquad N^{-1} = \begin{bmatrix} \mathbf{n}, \mathbf{n}^\perp \end{bmatrix}. \tag{3.4}$$

With this change of variables the Fourier transform of (3.3) is given by

$$
\begin{aligned}
\hat{I}(\mathbf{k}, \omega) &= \int\int\int I_0(y_1 - t v_n) e^{-i(\mathbf{k}^T \mathbf{n} y_1 + \mathbf{k}^T \mathbf{n}^\perp y_2 + \omega t)} \, dy_1 \, dy_2 \, dt \\
&= \hat{I}_0(\mathbf{k}^T \mathbf{n}) \int e^{-i \mathbf{k}^T \mathbf{n}^\perp y_2} \, dy_2 \int e^{-i \mathbf{k}^T \mathbf{n} v_n t} e^{-i \omega t} \, dt \\
&= \hat{I}_0(\mathbf{k}^T \mathbf{n}) \, \delta(\mathbf{k}^T \mathbf{n}^\perp) \, \delta(\mathbf{k}^T \mathbf{v}_n + \omega).
\end{aligned} \tag{3.5}
$$

From (3.5) it can be seen that all of the non-zero Fourier components associated with the moving profile must lie on a line through the origin in the frequency domain. The line is the intersection of the two planes $\omega = -\mathbf{v}^T \mathbf{k}$ and $\mathbf{k}^T \mathbf{n}^\perp = 0$. The speed v_n determines the angle between this line and the spatial frequency plane $\omega = 0$. The direction of motion $\mathbf{n}$ determines the orientation of the line about the ω-axis. Finally, there is a one-to-one mapping between lines intersecting the origin in frequency space and 2-d component velocities $\mathbf{v}_n$.

The aperture problem, and the relation between component velocity and 2-d velocity may be expressed in terms of the following facts:

– 2-d velocity maps to a plane through the origin in frequency space;
– the associated normal velocities correspond to lines in the plane; and
– an infinite number of planes contain a given line through the origin.

As the spatial structure of a translating signal becomes more one-dimensional, the non-zero power in the frequency domain becomes more concentrated about a single line (through the origin) in the plane.

3.3 Localization (the Uncertainty Relation)

As mentioned above, 2-d translation is used only as a local approximation to the time-varying behaviour of image intensity. But the above results, (3.2) and (3.5), assume that translation is constant over all of space-time. In this section we introduce a windowing function into the above analysis.

Let $W(\mathbf{x}, t)$ be a smooth window function, for which $|W(\mathbf{x})| \to 0$ as $\| \mathbf{x} \| \to \infty$, and let $I_W(\mathbf{x}, t; \mathbf{x}_0, t_0)$ denote the image neighbourhood about location $(\mathbf{x}_0, t_0)$; that is,

$$I_W(\mathbf{x}, t; \mathbf{x}_0, t_0) = W(\mathbf{x} - \mathbf{x}_0, t - t_0) \, I(\mathbf{x}, t). \tag{3.6}$$

From the convolution theorem (1.7), the effect of the window can be viewed in the frequency domain as the convolution of the respective Fourier transforms:

$$\mathcal{F}[I_W(\mathbf{x}, t; \mathbf{x}_0, t_0)] = \mathcal{F}[W(\mathbf{x} - \mathbf{x}_0, t - t_0)] \, * \, \mathcal{F}[I(\mathbf{x}, t)]$$
$$= \left(\hat{W}(\mathbf{k}, \omega) \, e^{-i(\mathbf{k}^T \mathbf{x}_0 + \omega t_0)} \right) \, * \, \hat{I}(\mathbf{x}, t) . \qquad (3.7)$$

Without loss of generality, assume that the window is placed at the origin so that $(\mathbf{x}_0, t_0) = (\mathbf{0}, 0)$.[1] Then, for smooth windows, (3.7) can be viewed as the application of a low-pass filter to the Fourier transform of the image, blurring its amplitude spectrum. In the case of image translation, the non-zero power will no longer lie strictly on a plane as in (3.2), or along a line as in (3.5); it now occupies a 3-d volume concentrated about the plane, or the line. As the extent (width) of the window decreases, there is greater blurring, and therefore poorer resolution in the frequency domain.

Another perspective from which to view this blurring of frequency space is provided by information theory [Gallager, 1968]. As the window size decreases, the blurring in frequency space implies a reduced ability to resolve different frequencies; that is, there is a loss in information. To see this, note that when a signal is smoothed by a band-limited filter, the number of samples necessary to represent the smoothed version of the signal changes in proportion to the bandwidth of the filter (the Sampling Theorem is discussed in Section 3.5). Therefore, as the window size in (3.6) decreases, so does the number of distinct Fourier coefficients that must be retained for a complete representation of the signal. This means that there is a limited number of degrees of freedom in a finite window of a band-limited function. As the window size decreases, so do the available degrees of freedom. In particular, less can be said about the frequency content of the signal, and fewer orientations and speeds can be resolved simultaneously.

This inability to achieve simultaneous resolution in space-time and in the frequency domain is well-known and is referred to as the *uncertainty relation* [Gabor, 1946; Slepian, 1983; Bracewell, 1978; Papoulis, 1968]; it provides a theoretical lower bound on the product of the spatial extent of a signal and the extent of its power spectrum. For a signal attaining this lower bound, a decrease in its spatial extent must be accompanied by an increase in the extent of its power spectrum.

The exact lower bound on the product of spatial extent and spectral extent depends on the definition of signal extent (or concentration); several definitions are discussed by Bracewell (1978) and Slepian (1983). The most

---

[1] Our goal is to consider the image structure in each neighbourhood relative to the centre of the neighbourhood rather than the nominal centre of the image. The phase shift in (3.7) simply marks the location of the window in terms of the nominal image centre. Of course, we could have also shifted the image centre, thereby inducing a simple phase shift to each harmonic in the image rather than the window.

common measure of extent, and hence the most common statement of the uncertainty relation, is given in terms of the variance of the power distribution of the signal. More precisely, for 1-d signals $f(x)$ with Fourier transforms $\hat{f}(k)$, the uncertainty relation states that

$$\sigma_x \sigma_k \geq \frac{1}{2},$$ (3.8)

where

$$\sigma_x^2 = \frac{\int x^2 |f(x)|^2 \, dx}{\int |f(x)|^2 \, dx}, \qquad \sigma_k^2 = \frac{\int k^2 |\hat{f}(k)|^2 \, dk}{\int |\hat{f}(k)|^2 \, dk}.$$ (3.9)

This can be proven using the Schwartz inequality in the form

$$4 \int |f(x)|^2 \, dx \int |g(x)|^2 \, dx \geq \left| \int (f^*(x)g(x) + f(x)g^*(x)) \, dx \right|^2$$ (3.10)

and Parseval's theorem (1.5) [Bracewell, 1978].

Our concern here is with multidimensional signals, and therefore we require a generalization of (3.8) to multiple dimensions. It should be expressed in terms of the product of the n-d extents (volumes) of the respective power distributions and, as above, it is natural to consider a measure of extent based on second-order moments. It is also natural to require that the measure of extent be rotation-invariant so that the volume of support is constant under orthogonal rotations of the coordinate system. Toward this end, for signals $f(\mathbf{x})$ where $\mathbf{x} \in \mathbb{R}^n$, with Fourier transforms $\hat{f}(\mathbf{k})$, the uncertainty relation can be generalized to

$$|C|^{\frac{1}{2}} |\hat{C}|^{\frac{1}{2}} \geq \frac{1}{2^n},$$ (3.11)

where C and $\hat{C}$ are the covariance matrices corresponding to $|f(\mathbf{x})|^2$ and $|\hat{f}(\mathbf{k})|^2$, and $|C|$ denotes the determinant of C. A proof for (3.11) is given in Appendix B.

Equality in (3.11), and hence in (3.8), is obtained with Gaussian functions:

$$G(\mathbf{x}; C) = \frac{1}{(2\pi)^{\frac{n}{2}} |C|^{\frac{1}{2}}} e^{-\frac{1}{2}\mathbf{x}^T C^{-1} \mathbf{x}},$$ (3.12)

the Fourier transforms of which, written $\hat{G}(\mathbf{k}; C)$, are unnormalized Gaussians with covariance matrices $\hat{C} = C^{-1}$; that is,

$$\hat{G}(\mathbf{k}; C) = e^{-\frac{1}{2}\mathbf{k}^T C \mathbf{k}}.$$ (3.13)

Accordingly, the covariance matrices of G^2 and $\hat{G}^2$ are given by $\frac{1}{2}C$ and $\frac{1}{2}C^{-1}$, the determinants of which are $2^{-n}|C|$ and $2^{-n}|C^{-1}| = 2^{-n}|C|^{-1}$. Thus, the product of their determinants achieves the lower bound in (3.11) on simultaneous localization in space and the frequency domain, which has been one reason cited for the use of Gaussian windows (e.g. [Marr and Hildreth, 1980]).

3.4   Multiple Velocities

The analysis in Sections 3.1 through 3.3 deals only with the simplest case of local translational motion. As outlined in Chapter 2, however, there may exist more than one legitimate velocity in a single spatiotemporal neighbourhood. Five common situations in which this occurs are:

- the virtual surfaces caused by specular reflection;
- the occlusion of one surface by another;
- translucency, the transmission of light through a surface;
- shadows under diffuse (ambient) illumination conditions; and
- atmospheric effects such as smoke in a room.

As discussed in Chapter 2, the combination of specular and diffuse components of reflection can be modelled linearly. This is convenient as the Fourier transform of the superposition of two signals is simply the superposition of their respective Fourier transforms. The superposition of two translating signals implies that power is concentrated on two planes through the origin in frequency space. They will be resolvable except on the line along which the two planes intersect. Of course, if we take the space-time aperture into account, the ability to resolve the two motions is a function of the signal structure, the window size, and the two velocities. So long as the local scales, orientations and component speeds are different, they should be resolvable even for smaller apertures. As the difference in scale, orientation, and speed between the two signals decreases, a single filter will begin to respond to both simultaneously, in which case resolution is not possible.

A second way in which more than one legitimate velocity can be said to occur in a single neighbourhood is the occlusion of one or more surfaces by another.[2] Consider a simple case involving a background with velocity $\mathbf{v}_b$ and an object moving with velocity $\mathbf{v}_o$. Let $I_o(\mathbf{x})$ and $I_b(\mathbf{x})$ denote the intensity profiles of the occluding foreground and the occluded background. For the occluding object we define a characteristic function that specifies the space occupied by the object:

$$\chi(\mathbf{x}) \;=\; \begin{cases} 1 & \text{if } I_o(\mathbf{x}) \neq 0 \\ 0 & \text{otherwise} \, . \end{cases} \tag{3.14}$$

The spatiotemporal intensity pattern resulting from the occluding object and the background can be expressed as

$$\begin{aligned} I(\mathbf{x}, t) &= I_o(\mathbf{x} - \mathbf{v}_o t) + [\, 1 - \chi(\mathbf{x} - \mathbf{v}_o t)\,]\, I_b(\mathbf{x} - \mathbf{v}_b t) \\ &= I_o(\mathbf{x} - \mathbf{v}_o t) + I_b(\mathbf{x} - \mathbf{v}_b t) - \chi(\mathbf{x} - \mathbf{v}_o t)\, I_b(\mathbf{x} - \mathbf{v}_b t) \, . \end{aligned} \tag{3.15}$$

---

[2]With small spatiotemporal apertures, relative to which objects are usually large, the occurrence of occlusion is sparse. Yet these boundaries are an important source of information, and a significant problem for optical flow techniques.

The Fourier transform of (3.15) is given by

$$
\begin{aligned}
\hat{I}(\mathbf{k}, \omega) = {} & \hat{I}_o(\mathbf{k})\delta(\omega + \mathbf{v}_o{}^T\mathbf{k}) \\
& + \hat{I}_b(\mathbf{k})\delta(\omega + \mathbf{v}_b{}^T\mathbf{k}) \\
& - [\hat{\chi}(\mathbf{k})\delta(\omega + \mathbf{v}_o{}^T\mathbf{k})] * [\hat{I}_b(\mathbf{k})\delta(\omega + \mathbf{v}_b{}^T\mathbf{k})] .
\end{aligned}
\tag{3.16}
$$

The first two terms correspond to the individual signals. The third term can be viewed as additive distortion caused by the occlusion of the background and the relative velocities of the foreground and background. Although a detailed examination of the expected forms of distortion is beyond the scope of this dissertation, several points are worth mentioning.

First, the background signal plays an important role in the resolution of the different velocities. If $I_b(\mathbf{x})$ has almost no power, or if it is uniform, in which case its Fourier transform is a delta function, then the second and third terms in (3.16) have no effect. If the only power in $I_b(\mathbf{x})$ is at low frequencies, then the second and third terms are somewhat more significant. In this case the third term will introduce power relatively close to the two planes $\omega = -\mathbf{v}_o{}^T\mathbf{k}$ and $\omega = -\mathbf{v}_b{}^T\mathbf{k}$, so that the concentration of power still reflects the two correct 2-d velocities. Trouble occurs when $I_b(\mathbf{x})$ or $\chi(\mathbf{x})$ contains significant power at high frequencies. In this case, the third term in (3.16) contains power at velocities far from $\mathbf{v}_b$ and $\mathbf{v}_o$, and incorrect measurements may occur. Orientation may also play a significant role. For example, occluding objects are generally larger than the aperture size, so that we may assume that the occluding boundary is straight-edged. In this case the characteristic function satisfies

$$
\chi(\mathbf{x}) = \begin{cases} 1 & \text{if } \mathbf{x}^T\mathbf{n} \geq 0 \\ 0 & \text{otherwise} \end{cases} , \quad \hat{\chi}(\mathbf{k}) = \pi\delta(\mathbf{k}) + i(\mathbf{k}^T\mathbf{n})^{-1} , \tag{3.17}
$$

where $\mathbf{n}$ is normal to the occluding edge. Therefore, $\hat{\chi}(\mathbf{k})$ contains power at high frequencies in some orientations, but power at only low frequencies in others. Although occlusion in the general case can spread power throughout frequency space, there are various situations in which the distribution of power will still reflect the correct velocities.

The third way in which more than one 2-d velocity might occur in the same neighbourhood is the transmission of light through translucent material. In most applications this does not occur as often as mirror-like reflections and occlusion, nor is it straightforward to deal with. Like the distortion caused by occlusion, the effects of translucency are nonlinear (e.g. [Whitted, 1980]). In addition, the transmission coefficient can be expected to change as a function of spatial position, an example of which is stained glass. As a simple model let the spatiotemporal intensity be given by

$$
I(\mathbf{x}, t) = \rho_0(\mathbf{x} - \mathbf{v}_0 t) I_1(\mathbf{x} - \mathbf{v}_1 t) , \tag{3.18}
$$

where ρ_0 reflects the density of the translucent material. The Fourier transform of (3.18) is

$$\hat{I}(\mathbf{k}, \omega) \;=\; \left(\hat{\rho}_0(\mathbf{k})\,\delta(\omega + \mathbf{v}_0{}^T\mathbf{k})\right) \, * \, \left(\hat{I}_1(\mathbf{k})\,\delta(\omega + \mathbf{v}_1{}^T\mathbf{k})\right) . \quad (3.19)$$

In this case, reliable measurements are expected only if one of the two signals has most of its power concentrated at low frequencies. If both signals have broad spectra, then there will be power spread throughout frequency space, far from the planes $\omega = -\mathbf{v}_0{}^T\mathbf{k}$ and $\omega = -\mathbf{v}_1{}^T\mathbf{k}$.

3.5　Temporal Sampling and Motion Blur

Finally, several comments are warranted about temporal sampling, smoothing (motion blur), and the relationship between scale and speed. These issues are important as they affect the information content of the input and the types of techniques that are permissible: for example, should we assume the input signal can be differentiated with respect to time?

The well-known sampling theorem [Dudgeon and Mersereau, 1986; Papoulis, 1968; Shannon and Weaver, 1963] states that:

> *A band-limited signal $f(t)$, with $\hat{f}(\omega) = 0$ for $\omega \geq 2\pi W$, can be determined (reconstructed) from its values $f(n\tau)$ at points $t = n\tau$, $n \in \mathbb{Z}$, provided that the sampling interval τ satisfies $\tau \leq 1/2W$.*

This suggests that, over duration T, such band-limited signals are completely represented by $2WT$ samples or degrees of freedom (for details see [Slepian, 1976]). In doing so, it defines the information content of smooth signals, and shows that efficient representations (encodings) are readily available along with straightforward methods for interpolation [Dudgeon and Mersereau, 1986; Schafer and Rabiner, 1973]. More precisely, from samples $f(n\tau)$ we can interpolate to find $f(t)$ as follows:

$$f(t) \;=\; \sum_{n=-\infty}^{\infty} f(n\tau)\, \frac{\sin\left(\frac{\pi}{\tau}(t - n\tau)\right)}{\frac{\pi}{\tau}(t - n\tau)} . \quad (3.20)$$

Similarly, (4.22) also provides a framework for numerical differentiation, since differentiating $f(t)$ amounts to simply differentiating the interpolant.

Sampling theory also concerns aliasing, the distortion in the representations of signals for which the sampling rate was too low to represent the highest frequencies. For a given sampling rate, the highest frequency that can be uniquely represented is referred to as the folding frequency. Higher frequencies are aliased: they appear indistinguishable from other frequencies below the folding frequency. In order to preserve the integrity (continuity) of

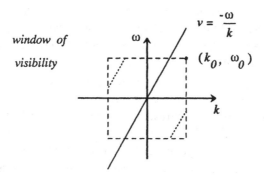

Figure 3.1. Window of Visibility: *After spatiotemporal smoothing and sampling, the highest spatial and temporal frequencies uniquely represented are given by k_0 and ω_0. Spatial and temporal frequencies associated with a signal translating with velocity v satisfy $v = -\omega/k$. The dotted lines denote aliased power that occurs within the window of visibility when the signal is not sufficiently band-limited prior to sampling.*

the represented signal, it is necessary to apply a low-pass filter to the signal *prior* to sampling, thereby removing all frequencies higher than the folding frequency. Then, the sampled encoding can be thought of as a complete representation of a smooth (analytic) signal (cf. [Slepian, 1976]). Following Watson and Ahumada (1983), the band of frequencies represented uniquely is referred to as the *window of visibility* (see Figure 3.1).

These standard results have several consequences. First, Section 3.1 showed that all the non-zero power associated with a translating pattern lies on a line through the origin in frequency space, where speed is given by the slope of the line $v = -\omega/k$. In order to ensure that the input is band-limited prior to sampling, spatiotemporal integration is necessary. From Figure 3.1, it is clear that temporal integration will remove high spatial frequencies moving at fast speeds; this phenomenon is often called *motion blur*. Although motion blur is sometimes thought to be a problem, it can also be viewed as the natural result of ensuring the integrity of the representation, and hence the ability to interpolate and differentiate. Similarly, the fact that fast speeds are only reliably measured at coarse scales is not surprising, since higher spatial frequencies should not exist for higher speeds. In fact, the window of visibility precisely specifies the relationship between speed and the available spatial frequencies. If the signal is not sufficiently band-limited prior to sampling, then non-zero (aliased) power will exist within the window of visibility at other velocities (shown as dotted lines in Figure 3.1).

One important consequence of these issues is that many of the motion sequences that are currently used to evaluate techniques for velocity measurement are severely corrupted by aliasing. The most popular method of creating image sequences, the results of which have been called *stop-and-shoot* sequences [Dutta et al., 1989], involves the digitization of a static scene with a stationary camera, from a sequence of different locations. Clearly, the effective motion of the camera does not affect the spatial resolution in each frame since no temporal integration is performed prior to temporal sampling. Interestingly, in standard cinematography (e.g. TV movies) it is common to *avoid* temporal smoothing in order to give viewers the impression of watching an actual, continuously varying, 3-d world. One explanation for this is that when viewers track (or stabilize) moving targets, they expect sufficient spatial detail to appear. If, on the other hand, viewers' eyes were stabilized to a physical point at the centre of the screen, then it is unlikely that they would notice motion blur, unless the temporal integration of the camera exceeded that of the human system.

In general, the amount of temporal integration should depend on the expected spectral content of the input signal and on the desired temporal sampling rate. It is also natural to assume a suspension system for the camera in order to avoid shocks that introduce significant amounts of power at high spatiotemporal frequencies.

Chapter 4

Velocity-Specific Representation

Low-pass and band-pass prefiltering are common precursors to many existing techniques for measuring image velocity (reviewed in Chapter 5). Depending on the particular approach, the main objectives of such filtering are noise reduction and/or the isolation of different types of image structure (e.g. zero-crossing contours, or different scales for coarse-fine analysis). More recently, it has been recognized that the filters themselves can be tuned to relatively narrow ranges of speed and orientation, as well as scale [Morgan, 1980; Fahle and Poggio, 1981; Adelson and Bergen, 1985; Fleet and Jepson, 1985; Watson and Ahumada, 1985; Heeger, 1988]. Since translating image patches can be viewed as oriented intensity structure in space-time (in much the same way as 1-d profiles can be oriented in space) they might therefore be detected using 3-d orientation-tuned filters. This intuition, in addition to the more formal analysis described in Chapter 3, leads to the design of *velocity-tuned filters*.

This chapter begins with a discussion of the basic design constraints for velocity-tuned filters, namely, localization in space-time, and localization in the frequency domain to ensure tuning to reasonably narrow ranges of scale, speed, and orientation. Section 4.2 discusses the construction of a family of filters that spans the necessary range of scales, speeds, and orientations to provide a useful representation of the raw image sequence. The representation, although not optimal, is reasonably efficient since the filter tunings do not overlap much and the filter outputs are subsampled; it is sufficient for the measurement of velocity that follows in Chapters 6 – 8. One requirement for this measurement technique, discussed in Appendix C, is the ability to numerically differentiate the subsampled representation of the filter outputs. Finally, Section 4.3 outlines the main advantages of velocity-specific representations.

4.1   Velocity-Tuned Filters

Chapter 3 showed that the non-zero power associated with a translating 1-d profile lies on a line through the origin in the frequency domain. The orientation and slope of the line depend continuously on the orientation and speed of the profile in space-time. Multiple component velocities will be resolvable at sufficiently high spatiotemporal frequencies (relative to the aperture size) where the power associated with each signal effectively occupies disjoint regions of frequency space. One of the simplest ways to detect (or resolve) them is with velocity-tuned filters, which respond strongly only to narrow ranges of speed, orientation and scale.

In discussing the design of velocity-tuned filters we begin with three desirable properties:

1. *Localization in Space-Time:* Our aim is to measure image velocity from the output of linear shift-invariant (convolution) operators of the form

$$R(\mathbf{x}_0, t_0) \;=\; \int\int K(\mathbf{x} - \mathbf{x}_0, t - t_0)\, I(\mathbf{x}, t)\, d\mathbf{x}\, dt \;. \qquad (4.1)$$

   To ensure that $R(\mathbf{x}_0, t_0)$ depends on local image structure, let $K(\mathbf{x}, t) = K_0(\mathbf{x}, t)\, W(\mathbf{x}, t)$ for some bounded function K_0, where $W(\mathbf{x}, t)$ is a window function ($|W(\mathbf{x}, t)| \rightarrow 0$ as $\| (\mathbf{x}, t) \| \rightarrow \infty$). Then, with $I_W(\mathbf{x}, t; \mathbf{x}_0, t_0) = W(\mathbf{x} - \mathbf{x}_0, t - t_0)\, I(\mathbf{x}, t)$, (4.1) can be rewritten as

$$R(\mathbf{x}_0, t_0) \;=\; \int\int K_0(\mathbf{x} - \mathbf{x}_0, t - t_0)\, I_W(\mathbf{x}, t; \mathbf{x}_0, t_0)\, d\mathbf{x}\, dt \;. \qquad (4.2)$$

   In other words, the filter can be thought of as a linear operator applied to the local signal structure in the neighbourhood about $(\mathbf{x}_0, t_0)^T$.

2. *Velocity Tuning:* The second condition concerns the tuning of the operator to orientation and speed. Sections 3.2 and 3.3 suggest that, for a filter to be selective to a relatively narrow range of orientations and speeds, its amplitude spectrum should fall mainly within a cone, whose central axis corresponds to the principal velocity to which the filter is tuned (see Figure 4.1). Given $\mathbf{v}_n = v_n\, \mathbf{n}$, the central axis is given by the unit vector

$$\mathbf{u} \;=\; \frac{(\mathbf{n}, -v_n)^T}{\sqrt{1 + v_n^2}} \;. \qquad (4.3)$$

   The opening angle of the cone determines the range of speeds and orientations to which the filter responds strongly. This constraint is discussed in standard multidimensional signal analysis in terms of beam-forming (e.g. [Dudgeon and Mersereau, 1984]).

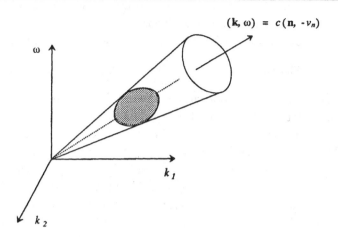

Figure 4.1. **Component Velocity Selectivity:** *This illustrates the amplitude spectrum of a linear filter that is tuned to component velocities near* $\mathbf{v}_n = v_n\,\mathbf{n}$. *The amplitude spectrum is concentrated in a cone, the central axis of which is given by* $(\mathbf{k}, \omega)^T = c\,(\mathbf{n}, -v_n)^T$ *for* $c \in \mathbb{R}$.

3. *Scale Specificity:* The extent of the filter support (the window in (4.2)) should depend on the scale of the image structure on which the measurements are based. This means, for example, that smaller windows should be used to measure fine scale structure. In terms of the amplitude spectra, this constraint means that the filters should be band-pass, with relatively narrow frequency pass-bands. Moreover, it is consonant with other arguments for scale-specific filters with constant octave bandwidths, often referred to as wavelets [Burt and Adelson, 1983; Koenderink, 1984; Crowley and Parker, 1984; Mallat, 1989].

As shown in Figure 4.1, Conditions 2 and 3 call for concentration of the amplitude spectrum in the frequency domain. Condition 1 calls for localization of the kernel in space-time. The theoretical limits on this joint resolution were discussed in Section 3.3, where it was shown that with one standard deviation as the measure of signal extent, Gaussian windows are optimally localized. This naturally leads to the class of 3-d Gabor kernels [Gabor, 1946]:[1]

$$Gabor(\mathbf{x}, t;\ \mathbf{k}_0, \omega_0, C) \;=\; e^{i(\mathbf{x}^T\mathbf{k}_0 + t\,\omega_0)}\, G(\mathbf{x}, t;\ C)\,, \qquad (4.4)$$

---

[1]Simultaneous localization in space-time and the frequency domain is not the most important precondition for a useful representation. Gabor functions, however, provide a convenient class of band-pass filters. They have a simple analytic form, and are easy and efficient to implement.

where $G(\mathbf{x}, t; C)$ denotes a 3-d Gaussian envelope with covariance matrix C (3.12). From the modulation theorem (1.9), and from (3.13), the Fourier transform of (4.4) is easily shown to be a Gaussian centred at $(\mathbf{k}_0, \omega_0)$:

$$\mathcal{F}[Gabor(\mathbf{k}, \omega; \mathbf{k}_0, \omega_0, C)] \;=\; \hat{G}(\mathbf{k} - \mathbf{k}_0, \omega - \omega_0; C). \qquad (4.5)$$

The Gaussian in (4.4) determines the envelope of the amplitude spectrum. The modulation of the complex exponential determines its placement in the frequency domain; $(\mathbf{k}_0, \omega_0)^T$ is called the peak tuning frequency of the filter.

The shape of the Gaussian window depends on C. In terms of velocity tuning, there are two cases of interest: *1)* the isotropic case in which C is diagonal, $C = \mathrm{diag}(\sigma, \sigma, \sigma)$; and *2)* the elliptical case in which the filter has non-zero aspect ratios and is therefore elongated in space-time as a function of the orientation and speed tuning of the filter. For example, a spatiotemporal Gabor function with aspect ratio $\alpha > 0$ can be defined using

$$(\mathbf{k}_0, \omega_0)^T \;=\; \frac{2\pi}{\lambda_0}(1, 0, 0)^T, \qquad\qquad C_0 = \mathrm{diag}(\sigma, \alpha\sigma, \alpha\sigma). \qquad (4.6)$$

It is tuned to stationary vertical structure with wavelengths near λ_0. Similar filters, tuned to the same spatiotemporal frequency band and component velocity $\mathbf{v}_n$, are characterized by covariance matrix C_j and modulation frequency $(\mathbf{k}_j, \omega_j)^T$, which are given by

$$(\mathbf{k}_j, \omega_j)^T \;=\; R_{\mathbf{u}}(\mathbf{k}_0, \omega_0)^T, \qquad\qquad C_j \;=\; R_{\mathbf{u}}^T C_0 R_{\mathbf{u}}. \qquad (4.7)$$

Here, $R_{\mathbf{u}}$ denotes an orthonormal rotation matrix that rotates $(1, 0, 0)^T$ onto $\mathbf{u}$ in (4.3).

In both cases the standard deviation σ determines the frequency bandwidth of the filter.[2] In particular, if the extent of the amplitude spectrum is defined as some proportion of the standard deviation in frequency space $\sigma_k = \sigma^{-1}$, then, for peak tuning frequency $f_0 = \| (\mathbf{k}_0, \omega_0) \| = 2\pi/\lambda_0$, the frequency bandwidth, in octaves, is defined as

$$\beta \;=\; \log_2 \left[\frac{f_0 + r\,\sigma_k}{f_0 - r\,\sigma_k} \right], \qquad (4.8)$$

where $r\sigma_k$ defines the frequency extent of the amplitude spectrum. Equivalently, given the frequency bandwidth β and the peak tuning frequency f_0, the standard deviation of the kernel, σ in (4.6), is easily shown to be

$$\sigma \;=\; \frac{r\,\lambda_0}{2\pi} \left(\frac{2^\beta + 1}{2^\beta - 1} \right). \qquad (4.9)$$

---

[2]The spatiotemporal frequency extent of the amplitude spectrum is defined as the width of the amplitude spectrum along the line through the origin of frequency space and the peak tuning frequency of the filter.

In general, the measure of extent r, the aspect ratio α, and the bandwidth β, collectively determine the number of different filters and the sampling rates for the filters' output.

4.2　A Velocity-Specific Representation

In constructing a family of filters to span the necessary range of scales, orientations and speeds, we aim for a pseudo scale-invariant family in which frequency space is decomposed into several scale-specific bands, each of which is decomposed in the same way into velocity-specific channels (cf. [Fleet and Jepson, 1989]). For simplicity we assume spherical amplitude spectra, so that $C = \mathrm{diag}(\sigma, \sigma, \sigma)$. Moreover, we assume that the spectral extent of the amplitude spectra is given by one standard deviation, $\sigma_k = 1/\sigma$, so that $r = 1$ in (4.8) and (4.9) (see Figure 4.2). The number of filters and their subsampling rates now depend mainly on the bandwidth β.

For convenience, consider a single spatiotemporal pass-band region, and assume that other scales are treated in a similar manner. Within this passband, we first consider the subset of filters that are tuned to zero temporal frequency (to speeds near zero), and span 180 degrees of spatial orientations. Following (4.9), for octave bandwidth β, and peak tuning frequency f_0, the standard deviation of the Gaussian envelope in frequency space is

$$\sigma_k = f_0 \left(\frac{2^\beta - 1}{2^\beta + 1} \right) . \tag{4.10}$$

It is convenient to have the different filters arranged so that the amplitude spectra of those tuned to neighbouring ranges of orientation will overlap (or just touch) at one standard deviation. This is illustrated in Figure 4.2. For reasonably narrow directional tuning, the range of orientations for which a single filter is responsible may be expressed as $2\Delta\theta \approx 2\sigma_k/f_0$. The appropriate number of differently tuned filter types is therefore

$$N(\beta) = \left\lceil \frac{\pi}{2\Delta\theta} \right\rceil \approx \left\lceil \frac{\pi \left(2^\beta + 1 \right)}{2 \left(2^\beta - 1 \right)} \right\rceil . \tag{4.11}$$

For example, with a bandwidth of 0.8 octaves this yields 6 filter types, each tuned to a range of $2\Delta\theta = 30$ degrees. The ceiling will cause a slightly greater amount of overlap when the bandwidth is such that the filters do not tile the pass-band region cleanly.

The same method can be used to continue decomposing the spatiotemporal frequency band into channels tuned to speeds other than zero. For example, given a collection of filters tuned to a speed of zero, and orientations at intervals of $2\Delta\theta = \sigma_k/f_0$, it is natural to then consider a set of filters

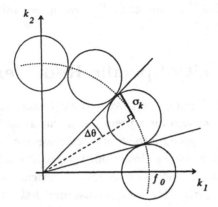

Figure 4.2. Decomposition into Orientation-Specific Channels: *Directionally tuned filters with neighbouring amplitude spectra should not overlap significantly. The orientation tuning $\Delta\theta$, measured at one standard deviation σ_k, is determined by the bandwidth β. Here, $\Delta\theta = \tan(\sigma_k/f_0)$. For small angles, with radius σ_k, it follows that $\Delta\theta \approx \sigma_k/f_0$.*

in the spatiotemporal frequency band centred at f_0 with frequency bandwidth β, tuned to speed $\tan(2\Delta\theta)$. That is, the filters tuned to different speeds should be spaced in the same way as those tuned to different orientations at zero speed, with their amplitude spectra just touching each other at one standard deviation. This is also clear from Figure 4.2 if the k_2-axis is replaced by the ω-axis.

A filter tuned to speed $\tan(2\Delta\theta)$ should be responsible for orientations in the approximate range of $2\Delta\theta/\cos(2\Delta\theta)$. To see this, note that, while σ_k remains constant for all filters in the spatiotemporal frequency band about f_0, the effective spatial frequency to which the filters at speed $\tan(2\Delta\theta)$ are tuned is $f_0\cos(2\Delta\theta)$. For a bandwidth of 0.8 octaves, this process eventually yields 23 filter types: 6 tuned to speeds about 0 with preferred directions at multiples of 30 degrees; 10 tuned to speeds of $1/\sqrt{3}$ with directions at every 36 degrees; 6 tuned to speeds of $\sqrt{3}$ with directions every 60 degrees; and a *flicker* channel tuned to non-zero temporal frequencies and zero spatial frequency [Fleet, 1988].

We now consider the discrete sampling of the output of a single filter type. As above, let $R(\mathbf{x}, t) = Gabor(\mathbf{x}, t; \mathbf{k}_0, \omega_0) * I(\mathbf{x}, t)$ be the output of a complex-valued Gabor kernel (4.4). For general images, $R(\mathbf{x}, t)$ is bandpass with its amplitude spectrum concentrated near $(\mathbf{k}_0, \omega_0)^T$. The minimal sampling rate for $R(\mathbf{x}, t)$ is defined according to the extent of the filter's amplitude spectrum (see Section 3.5 and Appendix C). Given the tiling of frequency space discussed above, it is sufficient to represent frequencies in the cube centred on $(\mathbf{k}_0, \omega_0)$, having sides of length $2\sigma_k$. The appropriate sampling rate is the Nyquist rate for frequency σ_k.[3] Therefore, the minimal sampling rate corresponds to a sampling distance of $\Delta s = \pi/\sigma_k = \pi\sigma$, where σ is the standard deviation of the Gaussian support in space-time. If the tiles do not overlap, then the total number of samples, at the minimal rate, will be equal to the number of pixels in the space-time region that are collectively encoded; that is, there are no redundant degrees of freedom in the representation.

Finally, it is important to note that the computation of component velocity described in Chapter 6 will require an interpolant for $R(\mathbf{x}, t)$. We will therefore sacrifice the efficiency of image encoding in favour of redundancy so that suitable interpolants are easier to compute. In particular, in the implementation reported in Chapter 7, we retain one (complex) sample every σ. This allows reasonably accurate interpolants to be obtained from the local samples of a single filter type. In Appendix C the numerical interpolation and differentiation of a subsampled version $R(\mathbf{x}, t)$ is explained in detail. There it is shown that the interpolation and differentiation can be performed by the convolution of a discrete kernel (with complex coefficients) with the subsampled filter output. The issue of accurate interpolation from a minimal sampling rate of $\Delta s = \pi\sigma$ is important, but is beyond the scope of this monograph.

4.3   Local Frequency-Based Approaches

As discussed in Chapter 5, most approaches to velocity computation make use of some form of spatial, scale-specific filtering. Others use spatiotemporal filtering, based on local filters tuned to scale, speed and orientation. We refer to such approaches as *frequency-based* to emphasize the Fourier analysis that serves as a basis for the velocity tuning. The main advantages of an initial velocity-specific representation of the image sequence are:

- *Noise Attenuation:*   Insofar as spatial prefiltering helps to lessen the effects of noise, the spatiotemporal nature of the velocity-tuned filters may provide further noise robustness because the filters can be designed to attenuate noise through time as well as space.

- *Spatiotemporal Localization:*   Like standard spatial filtering, the support kernels can be local in space-time. This avoids image structure

---

[3]This can be seen by demodulating the response $R(\mathbf{x}, t)$ by $\exp[-i(\mathbf{x}^T\mathbf{k}_0 + t\omega_0)]$, and considering the sampling rate for the cube now centred at the origin [Bracewell, 1978].

from spatially disjoint objects being merged accidentally into a single measurement. It also helps to ensure good spatial resolution of the velocity field, as it is not necessary to determine where within the window of support the measurement is relevant.

- *Velocity/Scale Resolution:* As discussed in Chapters 2 and 3, different physical processes are often manifested in images at different scales, with different velocities. The separation of signal structure according to scale, speed, and orientation that is provided by velocity-tuned filters therefore facilitates the resolution of different image properties that may reflect independent scene events. For example: the motion of specular reflections may be separated from the diffuse component of reflection; some occlusion relationships, such as those produced by waving your fingers in front of your eyes, are simplified; and some cases of translucency will be handled more reliably.

  The aperture problem occurs when only one dominant orientation exists in a given spatiotemporal window, in which case usually one filter will respond strongly. But, the occurrence of highly textured patterns in other neighbourhoods suggests that several filters may respond strongly. As discussed in Chapter 6, both cases can be handled, since the separation of image structure provided by the filters permits independent measurements of velocity at different orientations and scales within a single neighbourhood.

- *Efficiency:* Scale-specific and orientation-specific (also called subband) encodings have been shown to yield compression ratios of an order of magnitude or more [Burt and Adelson, 1983; Adelson et al., 1987]. We can expect even greater compression in space-time because of the high correlation that typically exists between frames.

Note that if the spatiotemporal support is kept small then, because of the uncertainty relation, we can expect only a limited separation of scales and velocities. The tuning will remain relatively crude. For example, each filter in the family of filters described above was responsible for a 30 degree range of orientations and angular speeds. In most vision applications a crude degree of separation is sufficient since more than two or three velocities in a small neighbourhood are unlikely to occur. Nevertheless, this also means that a subsequent stage of processing is needed because the accuracy required for tasks such as the determination of egomotion and surface parameters is greater than the tuning width of single filters (e.g. [Koenderink and van Doorn, 1987; Barron et al., 1991; Jepson and Heeger, 1990]). The use of velocity-tuned filters should be viewed as a stage of prefiltering, and not as the entire measurement process.

Chapter 5

Review of Existing Techniques

This chapter reviews four approaches to measuring image velocity:

- point-based differential methods;
- region-based matching methods;
- contour-based methods; and
- energy-based methods.

All these techniques extract local estimates of image translation, and in some cases the gradient of the the velocity field as well. As discussed in Section 3.2, localization in space-time is convenient for a variety of reasons, including the relevance of image translation, good resolution of the velocity field, the isolation of potentially independent scene properties and the isolation of the adverse effects of occlusion boundaries.

5.1 Differential Approaches

Differential techniques are often described as intensity-based in that velocity is computed directly from image intensity, in terms of its spatiotemporal derivatives. The first instances used only first-order image structure, and were originally based on an assumption that temporal variations in image intensity could be modelled locally by image translation [Fennema and Thompson, 1979; Horn and Schunck, 1981; Nagel, 1983, 1987; Tretiak and Pastor, 1984]; that is,

$$I(\mathbf{x}, t) = I(\mathbf{x} - \mathbf{v}\, t, 0) . \tag{5.1}$$

Following Horn and Schunck (1981), we can expand the right side of (5.1) as a Taylor series about $(\mathbf{x}, t)$ and discard the terms higher than first-order. This yields a constraint on the two components of $\mathbf{v}$ in terms of the partial derivatives of image intensity, called the *gradient constraint equation*:

$$I_x(\mathbf{x}, t)\, v_1 + I_y(\mathbf{x}, t)\, v_2 + I_t(\mathbf{x}, t) = 0 . \tag{5.2}$$

Interestingly, (5.2) also follows from an assumption of intensity conservation: $dI(\mathbf{x}, t)/dt = 0$. This implies that (5.2) is weaker than (5.1) since the conservation of intensity also admits a wide range of smooth deformations, including dilation and rotation. In other words, the motion constraint equation is appropriate for a wider class of temporal intensity variation than image translation.

In terms of explicit measurement, (5.2) yields the normal velocity of level contours of constant intensity, $\mathbf{v}_n = v_n \mathbf{n}$, where $\mathbf{n}$ is the direction of the spatial intensity gradient. The normal speed and direction are given by

$$v_n = \frac{-I_t(\mathbf{x}, t)}{\| (I_x(\mathbf{x}, t), I_y(\mathbf{x}, t)) \|} , \qquad (5.3a)$$

$$\mathbf{n} = \frac{(I_x(\mathbf{x}, t), I_y(\mathbf{x}, t))^T}{\| (I_x(\mathbf{x}, t), I_y(\mathbf{x}, t)) \|} . \qquad (5.3b)$$

One restriction on the use of (5.2) and (5.3) is that the two components of the 2-d velocity $\mathbf{v}$ are constrained by only one linear equation. Further constraints are therefore necessary to determine both elements of $\mathbf{v}$.

One approach is to exploit second-order derivatives of $I(\mathbf{x}, t)$ [Tretiak and Pastor, 1984; Nagel, 1983, 1987; Girosi et al., 1989; Uras et al., 1989]. For example, the following constraints can also be derived from (5.1):

$$\begin{bmatrix} I_{xx}(\mathbf{x}, t) & I_{yx}(\mathbf{x}, t) \\ I_{xy}(\mathbf{x}, t) & I_{yy}(\mathbf{x}, t) \end{bmatrix} \begin{pmatrix} v_1 \\ v_2 \end{pmatrix} + \begin{pmatrix} I_{tx}(\mathbf{x}, t) \\ I_{ty}(\mathbf{x}, t) \end{pmatrix} = \begin{pmatrix} 0 \\ 0 \end{pmatrix} \qquad (5.4)$$

Equation (5.4) also follows from the assumption that $\nabla I(\mathbf{x}, t)$ is conserved: $d\nabla I(\mathbf{x}, t)/dt = 0$. Strictly speaking, this means that no rotation or deformation is permitted, and hence (5.4) is a much stronger constraint than (5.2). In measuring image velocity, assuming that $d\nabla I(\mathbf{x}, t)/dt = 0$, the constraints in (5.4) may be used in isolation, or together with (5.2) and other differential constraints to yield an over-determined system of linear equations.

However, if the aperture problem prevails in a local neighbourhood (i.e. if the intensity function is effectively one-dimensional), then because of the sensitivity of numerical differentiation, we cannot expect to obtain second-order derivatives accurately enough to determine the tangential component of $\mathbf{v}$ reliably. As a consequence, the velocity estimates from second-order approaches are often sparse and noisy. Unreliable velocity measurements based on (5.4) can be detected using the condition number and the determinant of the Hessian matrix in (5.4) [Girosi et al., 1989; Nagel, 1983, 1987; Tretiak and Pastor, 1984]. The condition number (ideally near 1) is used to ensure that the system is far from singular, and the determinant (ideally large) is used to ensure that the second derivatives of I are large enough (in signal-to-noise terms). Of course, even if the size of the condition number precludes the full determination of $\mathbf{v}$, reasonable estimates of $\mathbf{v}_n$ may be readily accessible based solely on (5.3).

A second approach to introducing further constraints on the full 2-d velocity $\mathbf{v}$ is to combine measurements of component velocity in some spatial neighbourhood. Different measurements in local neighbourhoods can be used collectively to model the local 2-d velocity field, for example, as a low-order polynomial in v_1 and v_2, using a Hough transform or a least-squares fit [Fennema and Thompson, 1979; Glazer, 1981; Kearney et al., 1987; Waxman and Wohn, 1985]. Alternatively, Horn and Schunck (1981) suggested the use of a global *smoothness constraint* on $\mathbf{v}$. More precisely, the following energy function, defined over a domain of interest D, is minimized with respect to the velocity field:

$$\min_{\mathbf{v}} E = \int_D (I_x v_1 + I_y v_2 + I_t)^2 + \lambda^2 (\| \nabla v_1 \|^2 + \| \nabla v_2 \|^2) \, d\mathbf{x} \, , \ (5.5)$$

where the target velocity field $\mathbf{v}$ and the partial derivatives are functions of $\mathbf{x}$, and λ reflects the relative importance of smoothness in the solution. Nagel has since suggested the use of an *oriented-smoothness* assumption in which smoothness is not imposed across steep intensity gradients (edges) in an attempt to handle occlusion [Nagel, 1983, 1987; Nagel and Enkelmann, 1986].

In practice, the above techniques also involve some degree of prefiltering. Low-pass or band-pass prefiltering have been employed to improve the conditioning of numerical differentiation [Fennema and Thompson, 1979; Verri et al., 1989; Uras et al., 1989], and to allow for coarse-fine estimation (a control structure that combines measurements as well) [Glazer, 1987; Enkelmann, 1986; Nagel and Enkelmann, 1986]. When local explicit computation is used, low-pass spatial smoothing may also be applied to the individual components of $\mathbf{v}$ [Uras et al., 1989].

It is also interesting to note the simple relationship between equations (5.2) and (5.4) and the Fourier analysis in Chapter 3. From the derivative property of Fourier transforms (1.10), it is easy to show that the Fourier transform of motion constraint equation (5.2), with $\mathbf{v}$ held constant, is

$$i \, k_1 \, \hat{I}(\mathbf{k}, \omega) \, v_1 + i \, k_2 \, \hat{I}(\mathbf{k}, \omega) \, v_2 + i \omega \, \hat{I}(\mathbf{k}, \omega) = 0 \, . \qquad (5.6)$$

When $i \, \hat{I}(\mathbf{k}, \omega)$ is factored out of (5.6), what remains is the constraint plane $\omega = -\mathbf{v}^T \mathbf{k}$ as in (3.2). Second-order differential constraints also lead to the same plane. For example, the Fourier transform of the top row of (5.4) is

$$i \, k_1^2 \, \hat{I}(\mathbf{k}, \omega) \, v_1 + i \, k_2 k_1 \, \hat{I}(\mathbf{k}, \omega) \, v_2 + i \, k_1 \omega \, \hat{I}(\mathbf{k}, \omega) = 0 \, . \qquad (5.7)$$

As above, $i \, k_1 \, \hat{I}(\mathbf{k}, \omega)$ can be factored out, leaving $\omega = -\mathbf{v}^T \mathbf{k}$. Thus the frequency analysis is consistent with the intuition behind the use of differential techniques.

Finally, there are the common assumptions that gradient techniques can handle only shifts of less than 1 pixel/frame, and that intensity should be nearly linear in regions as large as expected displacements. These assumptions are necessitated by the use of only two frames, or by the existence of aliased power in (discrete) input signal. When only two frames are given, interpolation becomes problematic; we are limited to linear interpolation between the frames, and to first-order finite-difference approximations to temporal (and spatial) derivatives. Linear interpolation/differentiation will be accurate only when *1)* we have an inordinately high sampling rate [Gardenhire, 1964] or *2)* when the intensity structure is nearly linear. With more reasonable interpolants and numerical differentiation these restrictions no longer exist. Furthermore, note that the derivation of (5.2) from either (5.1) or $dI(\mathbf{x}, t)/dt = 0$ does not impose any restrictions on v_1 or v_2. But the existence of derivatives *is* assumed. This suggests that when there is aliasing in the input, the input signal is corrupted and numerical differentiation will not produce reasonable measurements. Aliasing will cease to be a problem when the displacements between frames are less than one pixel; that is, less than half the wavelength of the highest frequency in the signal.

5.2   Matching Approaches

Differential techniques are essentially point-based and they presuppose differentiability (to some order) of time-varying image intensity. If, however, accurate numerical differentiation is not possible because of noise, because only a small number of frames exist (e.g. 2), or because of severe aliasing in the image acquisition process (discussed in Section 3.5), then it is natural to turn to region-based matching techniques. Such approaches aim to find the best match between image regions in one frame with neighbouring regions in subsequent frames. It has been common to formulate the degree of fit based on some form of correlation measure, and hence these techniques are often referred to as correlation-based.

For instance, let $W(\mathbf{x})$ denote a 2-d window function centred at the origin with $|W(\mathbf{x})| \to 0$ as $\| \mathbf{x} \| \to \infty$. Then, $W(\mathbf{x}-\mathbf{x}_0) I(\mathbf{x}, t)$ denotes a windowed patch of the image centred at $\mathbf{x}_0$ at time t. The estimate of 2-d velocity $\tilde{\mathbf{v}}$ is defined as the shift $\mathbf{s}$ that yields the best fit between image regions centred at $\mathbf{x}_0$ at time t_0, and at $\mathbf{x}_0 + \mathbf{s}$ at time t_1; that is, between image patches

$$I_0(\mathbf{x}) \equiv W(\mathbf{x} - \mathbf{x}_0) I(\mathbf{x}, t_0) \, , \qquad\qquad (5.8\text{a})$$

$$I_1(\mathbf{x}; \mathbf{s}) \equiv W(\mathbf{x} - \mathbf{x}_0) I(\mathbf{x} + \mathbf{s}, t_1) \, . \qquad\qquad (5.8\text{b})$$

Finding the best match is essentially an optimization problem. For example, one might maximize a similarity measure (over $\mathbf{s}$), such as the normalized cross-correlation:

$$S(I_0(\mathbf{x}), I_1(\mathbf{x}; \mathbf{s})) = \frac{<I_0(\mathbf{x}), I_1(\mathbf{x}; \mathbf{s})>}{\parallel I_0(\mathbf{x}) \parallel \parallel I_1(\mathbf{x}; \mathbf{s}) \parallel}. \tag{5.9}$$

Or one might pose the problem in terms of the minimization of a distance measure, such as the sum-of-squared difference (SSD):

$$D(I_0(\mathbf{x}), I_1(\mathbf{x}; \mathbf{s})) = \parallel I_0(\mathbf{x}) - I_1(\mathbf{x}; \mathbf{s}) \parallel^2. \tag{5.10}$$

There is a close relationship between the SSD distance measure and both the cross-correlation similarity measure and differential techniques. First, note that minimizing the SSD distance amounts to maximizing the integral of cross-product term $2 < I_0(\mathbf{x}), I_1(\mathbf{x}; \mathbf{s}) >$. Second, the difference in (5.10) can be viewed as a first-order approximation to $\Delta t \, W(\mathbf{x} - \mathbf{x}_0) \, dI(\mathbf{x}, t)/dt$. Therefore, minimizing (5.10) yields an average (weighted) solution to the gradient constraint equation (5.2) over the window [Anandan, 1989]. If $W(\mathbf{x} - \mathbf{x}_0)$ is replaced by $\delta(\mathbf{x} - \mathbf{x}_0)$ then the resemblence is even clearer, because (5.10) then is equivalent to a first-order discrete-difference approximation to (5.2).

As with differential techniques, several other issues have proven important in practice. First, the aperture problem arises when the local intensity structure is effectively one-dimensional, in which case the local similarity (or distance) measure gives rise to surfaces (as a function of the shift $\mathbf{s}$) that are ridge-like, without a clearly defined maximum (or minimum). Anandan (1989) suggested that the conditioning with which 2-d velocity can be determined should be posed in terms of the curvature of the similarity (distance) surface about maxima (minima). Another thorny issue involves the large (inter-frame) displacement of periodic patterns and textured regions.[1] To handle these cases it is necessary to incorporate added constraints or some control structure. For example, Glazer et al. (1983) and Anandan (1989) employed initial prefiltering of the individual frames using band-pass filters, prior to the matching, and then a coarse-to-fine matching strategy in which displacement estimates at coarse scales were fed to finer levels as initial estimates. Band-pass prefiltering also helps remove low frequency signal structure which might otherwise dominate the similarity and distance measures. Smoothness constraints on the target velocity/displacement field have also been suggested such as that employed in (5.5) [Witkin et al., 1987; Anandan, 1989]. Finally, Little et al. (1988) simplified the measurement process by restricting velocities to integer numbers of pixels per frame. They replaced peak-finding by a winner-take-all voting rule, and thereby avoided the need for interpolating or over-sampling the local similarity (distance) surfaces.

---

[1]This is essentially a question of aliasing, as discussed in Section 3.5.

5.3   Contour Approaches

Contour-based approaches have appeared in essentially two flavours. In the first, referred to as *feature-based* approaches, image contours are explicitly detected and tracked from frame to frame. In the second, the contours are used to specify the locations in the image at which some form of differential technique should be applied.

Feature-based approaches were originally motivated by the belief that our perception of visual motion results from the identification of scene features in individual images (frames), followed by a correspondence (or tracking) process that associates features in one frame with those in the next. The use of contours and edges was attractive because they were thought to be relatively easy to extract and track. Moreover, it has been conjectured that they correspond to salient image properties, that they are stable through time, and that they are relatively high contrast image features with good signal-to-noise properties. Although these might be compelling arguments, the correspondence of edges to salient surface features remains untenable in any rigorous sense; the contours extracted with current edge-detection processes are rarely confined to a single smooth surface.

Edges can be extracted by one of several edge detection methods (e.g. [Marr and Hildreth, 1980; Canny 1986]), and provide a reasonably sparse collection of features from which image velocity is to be measured. Assuming that the edges move small distances relative to their density in the image, researchers have often considered contour correspondence to be a straightforward problem. From this correspondence, it has been common to define velocity in terms of the perpendicular distance from one contour to its corresponding contour in the next frame. Wu et al. (1989) employed iterative estimation to this basic approach to improve the accuracy of the estimates. Kass et al. (1988) tracked contours globally from frame to frame using a dynamic gradient-descent procedure. Although the motion of image curves has received considerable theoretical attention (e.g. [Waxman and Wohn, 1985; Bergholm, 1988; Faugeras, 1990]), the actual details of the contour tracking are often left unspecified.

In the second type of contour-based approach, edges are defined in terms of the output of a band-pass edge-enhancement operator. Velocity measurement, typically based on a differential method, is then restricted to edge locations and their local neighbourhoods. For example, edges are often defined as zero-crossings in the response to the Laplacian of a Gaussian convolved with the image; that is, locations satisfying

$$\nabla^2 G(\mathbf{x}; \sigma) * I(\mathbf{x}, t) \; = \; 0 \,, \tag{5.11}$$

where the directional derivative of the response normal to the zero-crossing contour should be non-zero. (Buxton and Buxton (1984) used a space-time extension of (5.11).) Along zero-crossing contours, it has been common to employ the gradient constraint equation leading to measurements of component velocity [Marr and Ullman, 1981; Hildreth, 1984; Duncan and Chou, 1988]. If at some point along a contour the radius of curvature is sufficiently small, then it is also possible to exploit second-order differential constraints [Gong, 1989].

Waxman et al. (1988) have argued that the correspondence problem is a significant obstacle for feature-based approaches. With respect to differential approaches, they argue that the time-varying behaviour of the band-pass filter output does not generally obey the assumption of constant amplitude which is required by differential techniques. As an alternative approach, they first create an edge map from the zero-crossing contours of $\nabla^2 G$ output. The edge map, with Dirac delta functions at the edge locations, is then convolved with a spatiotemporal Gaussian kernel to create a signal to which differential techniques could be applied. They used a second-order differential technique, as in (5.4), restricted to the initial edge locations.

There are several general problems with contour-based approaches, as well as specific problems with the approach suggested by Waxman et al. (1988). First, these approaches assume that features (including edge locations) can be well localized and are stable as a function of time. It is also implicitly assumed that the contours provide a rich description of the image so that some velocities do not go undetected. Finally, it is often assumed that the edges are well enough isolated that they facilitate a correspondence process. In the case of Waxman et al. (1988), the separation between edges must be larger, in some sense, than the scale of the Gaussian kernel that is convolved with the edge map. Hence, smaller profiles should be used for faster velocities. Conversely, the motion should be reasonably small relative to the scale of the kernel so that the convected profile can be differentiated.

Much of the recent work on contour-based motion analysis concerns the full determination of both components of **v** when component velocity measurements are given along the contour. As discussed in Section 3.1 there are explicit methods such as a least-squares fit to a parametric model [Waxman and Wohn, 1985], and iterative methods such as those based on smoothness assumptions [Hildreth, 1984; Gong, 1989].

5.4 Energy-Based Approaches

Energy-based methods are based on frequency analysis and an initial velocity-specific representation of the image sequence (see Chapters 3 and 4). They use the relative amplitudes of the outputs of differently tuned filters to compute image velocity [Adelson and Bergen, 1985, 1986; Heeger, 1987, 1988; Jahne, 1990; van Santen and Sperling, 1985; Grzywacz and Yuille, 1990; Shiwawa and Mase, 1990]. The different filters are used to sample the local power spectrum of the input. Remember that if the temporal variation of the input is due to local translation then all this power should be concentrated on the appropriate plane in frequency space.

To explain the basic approach, it is convenient to view the 3-d power spectrum of the spatiotemporal image as a distribution in $\mathbb{R}^3$ [Jahne, 1990; Shiwawa and Mase, 1990]. Let $P(\mathbf{k}) = |\hat{I}(\mathbf{k}, \omega)|^2$, where, for notational convenience $k_3 \equiv \omega$ in $P(\mathbf{k})$. The principal directions of the distribution are given by the eigenvectors of the covariance matrix, $C \in \mathbb{R}^{3 \times 3}$, the elements of which are the second-order moments given by

$$C_{ij} = \frac{\int \ldots \int k_i \, k_j \, P(\mathbf{k}) \, d\mathbf{k}}{\int \ldots \int P(\mathbf{k}) \, d\mathbf{k}}. \tag{5.12}$$

The eigenvectors form an orthonormal basis for frequency space. The eigenvalues specify the variance of the power spectrum in the corresponding principal directions. Then, in terms of velocity measurement,

- a translating 1-d profile will have all its power concentrated about a line through the origin, in which case only one eigenvalue should be significantly non-zero. The corresponding eigenvector coincides with the line that determines the component velocity.

- the translation of a 2-d textured image patch has its power concentrated about a plane in frequency space, and therefore two of the eigenvalues will be significantly non-zero. The corresponding eigenvectors will span the plane and therefore yield the 2-d velocity.

- for constant intensity all three eigenvalues will be zero, and for more complicated variations of spatiotemporal intensity, as in the case of multiple velocities or occlusion, all three eigenvalues will be non-zero.

Heeger (1987, 1988) suggested a nonlinear least-squares approach for the determination of 2-d velocity from the distribution of power over the differently tuned filters in each local neighbourhood of the image. The formulation was based on the expected magnitudes of response to a translating white noise pattern, which, from Parseval's Theorem, is given by the integral of the Gaussian amplitude spectrum and the plane along which the power spectrum for the noise is uniformly distributed.

More recently, Grzywacz and Yuille (1990) have extended Heeger's results to show that the correct velocity may be estimated directly from the relative magnitudes of filter outputs tuned to the same spatial frequencies

but different temporal frequencies. The importance of their results is that
1) velocities may be estimated from the output of a subset of the filters
rather than all filters; and *2)* the input power spectrum need not be flat as
in Heeger's approach.

An implicit assumption in these approaches is that velocity estimates
cannot be obtained from the output of a single filter. They compute velocity estimates based on the relative amplitudes of responses across differently
tuned filters; but in doing so they sacrifice velocity resolution, which is one
of the main advantages of local frequency-based approaches. Because of this,
two different component velocities could be confused with a single component
velocity; the sum of two different component velocities in a single neighbourhood, as can occur with textured, semi-transparent, or partially occluding
objects, can have the same distribution of output amplitudes as a single component velocity. This can occur even if the two component velocities lie in
the tuning regions of different filters. In addition, the robustness of these
approaches with respect to common deviations from image translation have
not been reported.

5.5 Discussion

5.5.1 Three-Stage Framework

Despite the differences in the above approaches, in practice they all fall within
a single conceptual framework consisting of three stages of processing:

1. *Prefiltering* – in which the image is filtered in order to extract image
   properties of interest and to reduce the effects of noise;

2. *Measurement* – in which constraints on the velocity field are extracted
   (e.g., derivatives of the filtered input, or peaks in a correlation surface);

3. *Integration* – in which the different constraints are combined, in conjunction with *a priori* constraints (such as smoothness), in order to
   obtain the final 2-d velocity field.

We now discuss the different techniques from this perspective (cf. [Burt et
al., 1983; Zucker and Iverson, 1987; Anandan, 1989]).

Prefiltering

Although prefiltering is not often discussed as a fundamental ingredient of
existing techniques for velocity measurement, it has proven to be an indis-
pensable component of successful algorithms. The most common form of

prefiltering has been convolution (linear and shift-invariant) with low-pass kernels (e.g. Gaussian windows) or band-pass kernels (e.g. Laplacian of a Gaussian). It is commonly applied spatially (to individual frames).

As images are generally thought to be noisy, one of the advantages of prefiltering is *noise reduction* [Fennema and Thompson, 1979]. This is particularly important with respect to differential approaches because of the noise sensitivity of numerical differentiation, which is required to estimate low-order partial derivatives of $I(\mathbf{x}, t)$. The use of low-pass smoothing in this context has been discussed in terms of regularization theory [Torre and Poggio, 1986; Bertero et al., 1988]. A second major advantage of prefiltering has been the enhancement of certain signal properties, such as specific scales and edges. For example, with respect to contour-based approaches, the edge detection procedure usually begins with edge-enhancing band-pass filtering. More generally, prefiltering can be used to separate different types of image structure, such as those at different *scales*, which may reflect different physical processes [Burt, 1981; Marr 1982; Crowley and Parker, 1984; Witkin, 1983]. With respect to velocity measurement, multi-scale representations have been used to permit the measurement of fast speeds at coarse spatial scales, and to allow for coarse-to-fine control strategies.

Band-pass filtering has also been an important first stage in matching approaches in order to remove mean illumination, which would otherwise dominate the similarity or distance measure, and to reduce effects of aliasing with textured patterns. Initial scale-specific filtering can be useful in selecting the scale of the signal structure in relation to the window size [Anandan, 1989]. This helps to ensure that there will be sufficient local structure to obtain a clear match, without losing too much spatial resolution.

Measurement

Constraints on image velocity are then extracted following the prefiltering. The different types of constraints have been discussed in Sections 5.1 – 5.4. Differential approaches compute the low-order partial derivatives of the intensity function, possibly at multiple scales. In matching methods a local maximum (minimum) is detected in a similarity (distance) measure as a function of local displacements. For contour approaches, edges are either tracked explicitly from frame to frame, or they serve as locations at which differential techniques are used.

Integration/Interpretation

The measurement techniques described above yield estimates of component velocity and/or 2-d velocity. In the case of component velocity it is clear that a subsequent stage of processing is required to determine 2-d velocity. Estimates of 2-d velocity, if extracted directly, are often sparse because of the aperture problem. Also, it is often assumed that, owing to the sensitivity of numerical differentiation, estimates of both component velocity and 2-d velocity will be noisy. As a consequence, it is generally believed that a subsequent stage of processing is necessary to

a) determine 2-d velocity when only component velocities are given,

b) determine 2-d velocity in areas without reliable (if any) measurements,

c) lessen the effects of noise.

A common objective in current approaches is the extraction of a smooth, unique, velocity field, throughout the entire image.

Two main approaches have been used to integrate local measurements. The first uses general smoothness assumptions, which leads to an iterative optimization (smoothing) process. Horn and Schunck (1981) defined smoothness in terms of the magnitude of the first-order variation of the two components of v (5.5). Nagel proposed an oriented-smoothness constraint in order to avoid smoothing across occlusion boundaries, and to respect the integrity of measurements at steep intensity gradients [Nagel, 1987; Nagel and Enkelmann, 1986]. Within a region-based matching framework Witkin et al. (1987) minimized (5.10) subject to the same smoothness constraint as in (5.5), while Anandan (1989) used an oriented-smoothness constraint like Nagel's. Hildreth (1984) minimized the first-order total variation of v as a function of arclength along contours. More recently, Yuille and Grzywacz (1989) have suggested the use of higher-order smoothness constraints on the estimated velocity field in order to limit the effective support of the smoothing, and to account for the integration of measurements from a number of different sources, such as component velocity estimates, full 2-d velocity estimates, and sparse, long-range feature correspondences. In addition to the theoretical results, they attempt to explain various perceptual phenomena such as motion capture [Ramachandran and Anstis, 1983; Ramachandran and Cavanagh, 1987], global coherence [Williams and Sekuler, 1984], and different manifestations of the aperture problem [Adelson and Movhson, 1982; Nakayama and Silverman, 1988a,b].

The other method used to combine neighbouring measurements involves the explicit computation of v in local regions in terms of predetermined parametric models, the most common of which assumes that the velocity field is constant over the region of interest (e.g. [Fennema and Thompson, 1979; Kearney et al., 1987; Little et al., 1988]). In this case there are two principal methods, namely, the Hough transform (or line clustering) and least-squares minimization. If local patches of the velocity field are modelled by linear or quadratic polynomials (e.g. [Glazer, 1981; Waxman and Wohn, 1985]), then care must be taken with the Hough transform to avoid an exponential increase in the storage required as a function of the dimensionality of the parameter space (e.g. [Xu et al., 1990]). In these cases, a straightforward approach is to use least-squares minimization. This form of explicit measurement does not produce 2-d image velocity estimates in regions that contain no primitive measurements as the iterative smoothness-based methods do.

When assessing these approaches, it is important to realize that the integration of primitive measurements is effectively a process of interpretation. When several measurements are merged into one, or equivalently, if one quantity is derived from several measurements, an implicit assumption is made that the primitive measurements reflect a coherent property of the world. In the context of velocity measurement, it is tacitly assumed in the above approaches that all local velocity measurements arise from the motion of a single smooth surface. As discussed in Chapter 2, such assumptions, like that of local image translation, do not adequately capture the expected spatiotemporal patterns of intensity that arise from typical projections of 3-d scenes. There are many instances in local neighbourhoods where the intensity variation reflects the motion of more than one surface (real or virtual) and the determination of whether a collection of measurements reflects the motion of only one surface or several surfaces is a matter of scene interpretation. In this light, the assumption of single smooth surface in each neighbourhood, or for each image contour, is overly restrictive. It also seems unlikely that this interpretation problem can be solved adequately in the image domain as a problem of smoothing [Yuille and Grzywacz, 1988], while 3-d effects such as occlusion are disregarded [Kolers, 1972].

It is also important to reconsider the filling in of those regions that are without measurements, and the reduction of measurement noise through smoothing. It can be argued that the fabrication of measurements where none occurred adds little reliable information. Moreover, given the accuracy required for subsequent tasks that use estimates of the velocity field, such as the determination of egomotion or surface parameters, it seems unlikely that the fabricated measurements will be sufficient. With respect to the problem of sparse, incorrect, and noisy velocity estimates, two approaches are readily apparent. We could either determine a reasonable model for the errors (which is unlikely to be Gaussian) that could be used by subsequent processes in a principled way, or we could investigate more reliable measurement techniques.

Despite the smooth appearance of flow fields that result from the common forms of smoothing, the important issue is the accuracy of the measurements, that is, how well they approximate the motion field. The approach described in this monograph stems from an attempt to derive a more robust and accurate measurement technique that yields a dense set of estimates.

5.5.2   Summary

As discussed throughout Chapter 2, our goal is to define an image property that, when tracked through space-time, yields reliable measurements of the projected motion field of some surface. The two principal questions are: What image property should be used? And how should the tracking be accomplished?

The approaches described above implicitly tracked one of: *1)* the intensity of raw images or filtered versions thereof (as in the component of velocity normal to level contours of constant intensity); *2)* image contours (commonly defined as level zero-crossings of band-pass filtered images), or *3)* patterns of intensity (raw or filtered images), as in region-based matching approaches. All these approaches work well given image translation, but they sometimes perform poorly under general viewing conditions. In particular, they are sensitive to contrast variations, and to image deformation such as dilation or shear. Also common to the above approaches is a restriction to a single velocity estimate at any given image location. Such a restriction is implicit in the smoothness assumptions, the combination of local estimates to compute 2-d velocity from component velocities, and in the use of coarse-to-fine control strategies that assume that the same physical structure is reflected coherently across a wide range of scales.

Part II

Phase-Based Velocity Measurement

Part II of the monograph addresses the quantitative measurement of velocity in image sequences, and proposes a new approach to the problem. As discussed in Part I, the important issues are

- the accuracy with which velocity can be computed,

- robustness with respect to smooth contrast variations and geometric deformation; that is, deviations from image translation that are typical in projections of 3-d scenes,

- localization in space-time,

- noise robustness, and

- the ability to discern different velocities within a single neighbourhood, for example, due to occlusion, transparency, shadows, specular reflections, or atmospheric effects.

Our approach is based on the use of a collection of velocity-tuned filters, and the phase behaviour of their respective outputs. In particular, the phase behaviour is used to extract measurements of *component velocity*: the component of 2-d velocity in the direction normal to oriented structure in the image (defined in Chapter 6 in terms of level phase contours).

The use of phase information offers several advantages over the techniques reviewed in Chapter 5 with respect to some of the issues listed above. In particular, in the experiments reported below we find that the measurement accuracy significantly exceeds the tuning width of single filters by an order of magnitude. This is obtained without explicit sub-pixel feature localization. Furthermore, the support required for the measurements is only marginally larger than that of the initial velocity-tuned filters, and because the measurements are computed from the responses of similarly-tuned filters, velocity resolution is preserved. Finally, as discussed at length in Part III, phase information is robust with respect to typical deviations from image translation. In particular, it is more robust than amplitude with respect to changes in contrast, scale, orientation, and speed.

The concentration on component velocity stems from a desire for local measurements, and from the occurrence of the aperture problem, in which case *only* normal components of 2-d velocity can be measured reliably. In previous approaches, the computation of *full* 2-d velocity, the interpolation of the velocity field over regions without measurements, and the reduction of noise effects have necessitated assumptions such as smoothness and uniqueness of the velocity field, as well as the use of relatively larger regions of measurement support (compared to the initial filters). As discussed in Section 5.5, *a priori* assumptions such as smoothness and uniqueness are too restrictive for image sequences of general scenes. Moreover, the integration of measurements over large domains makes the significant leap of faith that all the measurements arise from the motion of the same smooth surface in the scene. In considering just normal components of velocity we obtain more accurate estimates of motion within smaller apertures, which leads to better spatial resolution of velocity fields. As a result, the effects of image rotation and perspective distortions such as shear and dilation, as well as measurements near occlusion boundaries, may be handled more reliably.

Chapter 6 introduces the notion of phase information, discusses its stability properties, and suggests a phase-based definition for component velocity which is then used to design a measurement technique. We then outline the similarities of this approach to the alternative notions of image velocity described in Chapter 5. Chapter 7 reports an extensive series of experiments with the new approach, and Chapter 8 extends the experimental work by solving for the full 2-d velocity using a straightforward method, further demonstrating the accuracy and robustness of the component velocity estimates. The work discussed in these chapters was originally reported in [Fleet and Jepson, 1990].

Chapter 6

Image Velocity as Local Phase Behaviour

This chapter is intended to motivate and introduce a phase-based definition of image velocity, and a corresponding technique for its measurement. To begin, let $R(\mathbf{x}, t)$ denote the response of a velocity-tuned band-pass filter. Because the filter kernels, such as $Gabor(\mathbf{x}, t; \mathbf{k}_0, \omega_0, C)$ are complex-valued, $R(\mathbf{x}, t)$ is also complex-valued and can therefore be expressed as

$$R(\mathbf{x}, t) \;=\; \rho(\mathbf{x}, t)\, e^{i\phi(\mathbf{x}, t)}, \tag{6.1}$$

where $\rho(\mathbf{x}, t)$ and $\phi(\mathbf{x}, t)$ denote its amplitude and phase components (1.4):

$$\rho(\mathbf{x}, t) \;=\; |R(\mathbf{x}, t)| \,,$$
$$\phi(\mathbf{x}, t) \;=\; \arg[R(\mathbf{x}, t)] \,.$$

In the search for an appropriate definition of image velocity, our goal is to determine which properties of $R(\mathbf{x}, t)$ evolve in time according to the motion field. The thesis set forth in this monograph purports that the phase response $\phi(\mathbf{x}, t)$ is such a property. In particular, it is shown that the temporal evolution of (spatial) contours of constant phase provides a better approximation to the motion field than do contours of constant amplitude $\rho(\mathbf{x}, t)$ or the level contours of $R(\mathbf{x}, t)$.

We begin with a demonstration of the robustness of $\phi(\mathbf{x}, t)$ as compared with $\rho(\mathbf{x}, t)$ in Section 6.1. Section 6.2 then introduces the phase-based definition of component velocity. Section 6.3 shows the connection between this definition and local frequency analysis, and Section 6.4 discusses its relationship to the definitions implicit in other techniques.

6.1   Why Phase Information?

If the temporal variation of image intensity was due solely to image translation, as in $I(\mathbf{x}, t) = I(\mathbf{x} - \mathbf{v}t, 0)$, then it is easy to show that the filter outputs would also translate, as in $R(\mathbf{x}, t) = R(\mathbf{x} - \mathbf{v}t, 0)$. Moreover, we would therefore expect that, in principle, all the techniques described in Chapter 5 should produce accurate estimate of the velocity $\mathbf{v}$, so long as the aperture problem and undersampling of the signal are not too severe.

However, image translation is only a rough approximation to the typical time-varying behaviour of image intensity. As discussed in Chapter 2, a more realistic model includes contrast variation and geometric deformation due to perspective projection; and it is from this standpoint that we propose the use of phase information. In particular, we argue that the evolution of phase contours provides a better approximation to the projected motion field than the filter response $R(\mathbf{x}, t)$, in that the amplitude of response $\rho(\mathbf{x}, t)$ is generally very sensitive to changes in contrast and to local variations in the scale, speed, and orientation of the input.

We demonstrate the robustness of phase compared to amplitude using several 1-d examples which serve to approximate the dilation of an image as a camera approaches a planar surface. In particular, we consider the time-varying image

$$I(x, t) \;\; = \;\; I(x\,(1 - \alpha t), 0)\,, \tag{6.2}$$

for some $\alpha > 0$. The initial pattern $I_0(x) \equiv I(x, 0)$ is simply stretched as t increases. The velocity field for this deformation is given by the motion of fixed points, say ξ, in the 1-d pattern $I_0(x)$. In image coordinates these points appear on paths generated by $x\,(1 - \alpha t) = \xi$. These paths are clearly visible from the inputs in Figures 6.1 and 6.2 (*top-left*).

Figure 6.1 (*top*) shows the time-varying intensity pattern generated by (6.2) for $I_0(x) = \sin(2\pi f_0 x)$, and the time-varying response of the real part of a Gabor filter (4.4) tuned to spatial frequency $2\pi f_0$ and to zero velocity (vertically oriented structure in Figure 6.1). The amplitude and phase components of $R(x, t)$ are shown, as functions of space and time, in Figure 6.1 (*middle*). Finally, the bottom panels show the level contours of constant amplitude and constant phase superimposed upon the input. While the phase contours provide a good approximation to the motion field, the amplitude contours do not. The reasons for this amplitude behaviour are straightforward: The behaviour of $\rho(\mathbf{x}, t)$ reflects how well the local input structure matches the filter tuning. In this case, with the amplitude of the input sinusoid held constant, they match very well in the centre of the image and less so as one moves out from the centre. Other things being equal, $\rho(x, t)$ increases for inputs closer to the principal frequency to which the filter is

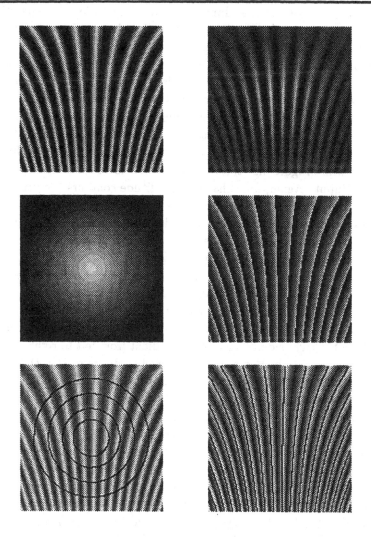

Figure 6.1. **Dilation of Sinusoid:** *(top-left)* $I(x,t) = \sin(2\pi x f_0(1 - \alpha t))$ *where* $f_0 = 1/12.5$ *pixels, and the origin,* $x = t = 0$, *is in the centre. The image width is 150 pixels, and* $\alpha = 0.0005$. *Time is shown on the vertical axis and space on the horizontal. The filter was tuned to speeds about 0, frequencies about* f_0, *and a bandwidth of 0.8 octaves. (top-right) Real part of Gabor output. (middle) Amplitude and phase outputs. (bottom) Level contours of constant amplitude and phase superimposed on the input.*

tuned, because the filter's amplitude spectrum is Gaussian-shaped. Therefore, it should be evident that $\rho(x, t)$ depends on the local scale and speed
of the input. In two spatial dimensions the amplitude will depend on the
orientation of the input as well as speed and scale, all of which vary locally
in typical projections of 3-d scenes.

Figure 6.2 depicts similar camera movement, but $I_0(x)$ is now taken to
be a sample of white Gaussian noise. This is a more realistic example in that
$I_0(x)$ now has structure at all scales, so that different image structure will be
emphasized by the filter at different times at different locations. Figure 6.2
(*top*) shows the input, and the real part of the Gabor response. The time-
varying amplitude and phase components of response are given in Figure 6.2
(*middle*), while Figure 6.2 (*bottom*) shows their level contours superimposed
upon the input. Again, note that the amplitude contours are very sensitive
to small scale perturbations, and do not evolve according to the motion field.
On the other hand, the phase contours do provide a good approximation to
the motion field except near a few spatiotemporal locations. These exceptions
lead to poor measurements and are discussed in Chapter 10.

Similar simulations show that phase behaviour is relatively insensitive to
contrast and shading variations. As a crude example, the top two images in
Figure 6.3 show an input pattern similar to that in Figure 6.2, and a shaded
version of it. The shading function is constant up the left side, and varies dark
to light on the right side. The shading gradient from left to right is constant.
The bottom images in Figure 6.3 show level contours of $\rho(x, t)$ and $\phi(x, t)$
superimposed upon the unshaded version of the input. The shading gradient
primarily affects the amplitude of the filter output, while the phase signal
still evolves according to the input motion field.

In general, as long as the photometric effects (e.g. shadows, highlights,
etc.) do not introduce power over a wide frequency band relative to the
surface texture, the phase behaviour of most filter outputs will be largely
unaffected. For example, in the case of smooth shading gradients, the main
spatiotemporal photometric effects are relatively smooth contrast variations
(cf. Section 2.2.3). Although the amplitude responses of filters tuned to
high spatial frequencies will be affected, the phase behaviour will remain
stable. In the neighbourhood of a steep shading gradient (e.g. a shadow
boundary) the spatiotemporal phase structure will reflect the motion of the
shading edge. We expect that for textured surfaces, the phase behaviour
will be predominately influenced by the projected motion field. Conversely,
specular flashes and complex textures caused by self-shadowing will cause
problems.

As a final demonstration, Figure 6.4 shows a dilating noise signal (*top-
left*), and the same signal superimposed onto a spatially-varying, temporal
flicker pattern (*top-right*). This serves as a crude example of two simultaneous

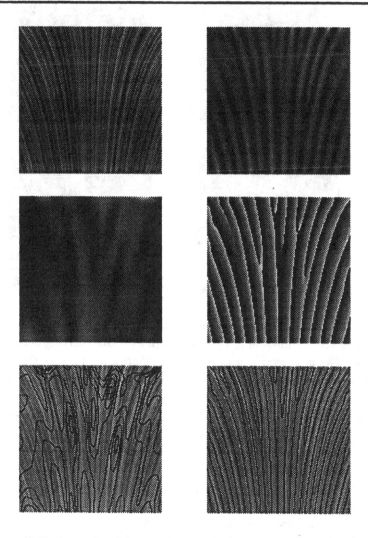

Figure 6.2. **Dilation of White Noise Pattern:** *The camera approaches a surface textured with white Gaussian noise. The filter tuning was identical to that in Figure 6.1. (top-right) Real part of Gabor output. (middle) Amplitude and phase outputs. (bottom) Level contours of constant amplitude and phase superimposed on the input.*

Figure 6.3. **Shaded Dilation Pattern:** *The camera approaches a surface covered by white Gaussian noise, modulated by a time-varying linear shading gradient. The filter tuning was identical to that in Figure 6.1. (top) The raw expanding input signal and the shaded version. (bottom) Level contours of constant amplitude and phase superimposed on the unshaded input.*

oriented patterns, as in the case of specular reflections. The filter that was applied to the superposition of the two signals was identical to that used in previous figures, and is evidently not disturbed by the flicker. This is clear from Figure 6.4 (*bottom*) where level contours of $\rho(x, t)$ and $\phi(x, t)$ are shown, superimposed upon the uncorrupted input.

With this preliminary motivation for the use of phase information, the basic ideas behind the approach can be summarized. First a family of velocity-tuned band-pass filters is applied to the image sequence. Second, image velocity is defined in the output of each filter independently as the temporal evolution of level phase contours. Toward this end, a simple threshold technique is used to detect and remove velocity measurements in regions for which phase contours are not likely to provide reliable information about the motion field. The performance of the resulting technique is evaluated through extensive experimentation in Chapters 7 and 8.

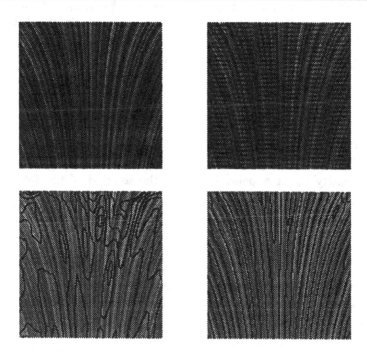

Figure 6.4. **Superposition of Noise Pattern and Temporal Flicker:** *(top-left) Dilating Gaussian noise pattern. (top-right) Added to the noise pattern is the product of a spatial sinusoid (wavelength of 50 pixels) and a temporal sinusoid (wavelength of 4.5 pixels). The amplitude of the modulated sinusoid is equal to the standard deviation of the noise. (bottom) Level contours of $\rho(x, t)$ and $\phi(x, t)$ superimposed upon the raw noise signal.*

6.2 Component Velocity from Level Phase Contours

With the decision to track phase information, we define image velocity in terms of space-time surfaces of constant phase; that is, solutions to

$$\phi(\mathbf{x}, t) = c, \qquad c \in \mathbb{R}. \tag{6.3}$$

Assuming that constant phase surfaces evolve according to the motion field, a point $\mathbf{x}_0(t)$ moving with the motion field satisfies $\phi(\mathbf{x}_0(t), t) = c$. Differentiating with respect to t, we find that

$$\phi_{\mathbf{x}}(\mathbf{x}, t)^T \mathbf{v} + \phi_t(\mathbf{x}, t) = 0, \tag{6.4}$$

where $\phi_\mathbf{x}(\mathbf{x}, t) = (\phi_x(\mathbf{x}, t), \phi_y(\mathbf{x}, t))^T$, and $\mathbf{v} = \frac{d}{dt}(x_0, y_0)^T$. Like the gradient constraint equation introduced by Horn and Schunck (1981), (6.4) only constrains the component of $\mathbf{v}$ in the direction normal to spatial contours of constant phase, that is, in the direction of the spatial gradient

$$\mathbf{n}(\mathbf{x}, t) = \frac{\phi_\mathbf{x}(\mathbf{x}, t)}{\| \phi_\mathbf{x}(\mathbf{x}, t) \|} . \tag{6.5}$$

The component of $\mathbf{v}$ in the direction $\phi_\mathbf{x}(\mathbf{x}, t)^\perp$ is not determined.

The combination of (6.4) with (6.5) provides our definition for component image velocity $\mathbf{v}_n$ at a point $(\mathbf{x}, t)$, as the solution of the following two equations:

$$\phi_\mathbf{x}(\mathbf{x}, t)^T \mathbf{v}_n + \phi_t(\mathbf{x}, t) = 0 , \tag{6.6a}$$

$$\mathbf{v}_n = \alpha \, \mathbf{n}(\mathbf{x}, t) , \quad \alpha \in \mathbb{R} . \tag{6.6b}$$

After outlining the computation of $\nabla\phi(\mathbf{x}, t)$ in the following section, we discuss this definition in relation to previously used definitions, to frequency analysis, and to phase-difference methods used in stereo.

6.2.1   Measurement of Phase Gradient

Rather than compute $\nabla\phi(\mathbf{x}, t)$ from the subsampled phase signal directly, it is more convenient to use the identity

$$\nabla\phi(\mathbf{x}, t) = \frac{\text{Im}[R^*(\mathbf{x}, t) \, \nabla R(\mathbf{x}, t)]}{\rho^2(\mathbf{x}, t)} , \tag{6.7}$$

where $\text{Im}[\mathbf{z}] \equiv (\text{Im}[z_1], \text{Im}[z_2], \text{Im}[z_3])^T$. In terms of the real and imaginary parts of $R(\mathbf{x}, t)$ and $\nabla R(\mathbf{x}, t)$, (6.7) becomes

$$\nabla\phi(\mathbf{x}, t) = \frac{\text{Im}[\nabla R(\mathbf{x}, t)] \, \text{Re}[R(\mathbf{x}, t)] - \text{Re}[\nabla R(\mathbf{x}, t)] \, \text{Im}[R(\mathbf{x}, t)]}{\text{Re}[R(\mathbf{x}, t)]^2 + \text{Im}[R(\mathbf{x}, t)]^2} . \tag{6.8}$$

In the complex plane, as depicted in Figure 6.5, (6.8) corresponds to a projection of each component of $\nabla R(\mathbf{x}, t)$ onto the unit vector orthogonal to $R(\mathbf{x}, t)$. This formulation eliminates the need for an explicit trigonometric function to compute the phase signal from $R(\mathbf{x}, t)$. It also avoids problems arising from phase wrapping/unwrapping and discontinuities.

In order to compute the phase gradient from (6.8), it is sufficient to have $R(\mathbf{x}, t)$ and its gradient $\nabla R(\mathbf{x}, t)$. The numerical interpolation and differentiation of $R(\mathbf{x}, t)$ is discussed in Appendix C.

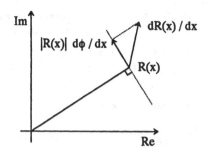

Figure 6.5. **Computation of** $\nabla\phi(\mathbf{x}, t)$**:** *In geometric terms, the phase derivative $d\phi(x)/dx$ is equivalent to $dR(x)/dx$ projected onto the unit vector perpendicular to $R(x)$, then scaled by $1/|R(x)|$.*

6.3 Local Frequency Analysis

We now examine the relationship between the phase gradient and frequency analysis. To begin, consider a simplified 2-d situation in which the input is a *uniform* sinusoidal waveform: $I(\mathbf{x}) = \cos(\mathbf{k}_1^T\mathbf{x})$. It is easily shown that the output $R(\mathbf{x})$ of a Gabor filter tuned to $\mathbf{k}_0$ (near $\mathbf{k}_1$) is a complex waveform with frequency $\mathbf{k}_1$ and amplitude $\hat{G}(\mathbf{k}_0 - \mathbf{k}_1; C)$.[1] The output phase $\phi(\mathbf{x}) = \mathbf{k}_1^T\mathbf{x}$, like the phase of the input, is a linear function of $\mathbf{x}$. The spatial phase gradient is equal to the frequency $\mathbf{k}_1$, which specifies the directional information. Similarly, for a sinusoidal plane-wave in space-time $\rho e^{i\phi(\mathbf{x}, t)}$, where $\phi(\mathbf{x}, t) = \mathbf{k}^T\mathbf{x} + \omega t$, the phase gradient yields the spatial and temporal frequencies (i.e. $\nabla\phi(\mathbf{x}, t) = (\mathbf{k}, \omega)^T$).

The present situation is somewhat more involved, for the Gabor output is rarely a sinusoidal waveform. Rather, it is a *nonuniform* waveform because its amplitude is not constant through space-time, nor does it have linear phase. Despite this, the phase gradient can still be used to define the instantaneous spatial and temporal frequencies of the output signal [Papoulis, 1965]. To understand this, note that $R(\mathbf{x}, t)$ will generally have its power concentrated about the filter tuning $(\mathbf{k}_0, \omega_0)^T$. Using the modulation property (1.9) we could demodulate $R(\mathbf{x}, t)$ to create a signal $M(\mathbf{x}, t)$, a demodulated version of $R(\mathbf{x}, t)$ that has its power concentrated about the origin in frequency space. In other words, from (6.1), $R(\mathbf{x}, t)$ may be rewritten as

$$R(\mathbf{x}, t) = \rho(\mathbf{x}, t) \, e^{i\left[\phi_M(\mathbf{x}, t) + (\mathbf{x}^T\mathbf{k}_0 + t\omega_0)\right]}, \qquad (6.9)$$

where $M(\mathbf{x}, t) \equiv \rho(\mathbf{x}, t) \, e^{i\phi_M(\mathbf{x}, t)}$ is a low-pass signal. This suggests that

---

[1]This assumes that $\mathbf{k}_0 \approx \mathbf{k}_1$ so that the filter's response to the input harmonic at frequency $-\mathbf{k}_1$ is negligible. Alternatively, let the input be $\exp(i\mathbf{k}_1^T\mathbf{x})$.

we view $R(\mathbf{x}, t)$ as a slowly varying modulation of the base signal to which the filter was tuned $e^{i(\mathbf{x}^T \mathbf{k}_0 + t\omega)}$. The phase behaviour of $M(\mathbf{x}, t)$ can be viewed as a slowly varying correction to the linear phase behaviour of the base signal. We therefore expect that the phase of $R(\mathbf{x}, t)$ should be nearly linear,[2] and if the amplitude is relatively flat, then $R(\mathbf{x}, t)$ is expected to be nearly sinusoidal. Finally, again following Papoulis (1965), the local (instantaneous) frequency is defined as the phase gradient

$$(\mathbf{k}(\mathbf{x}, t),\ \omega(\mathbf{x}, t))^T \quad \equiv \quad \nabla \phi(\mathbf{x}, t) . \tag{6.10}$$

If the phase of $M(\mathbf{x}, t)$ is linear in space-time, such as $\phi_M(x) = \mathbf{k}_1^T \mathbf{x} + \omega_1 t$, then $R(\mathbf{x}, t)$ is just an amplitude-modulated sinusoid with constant frequency $(\mathbf{k}_1 + \mathbf{k}_0, \omega_1 + \omega_0)^T$. Otherwise, the phase gradient $\nabla \phi(\mathbf{x}, t) = \nabla \phi_M(\mathbf{x}, t) + (\mathbf{k}_0, \omega_0)^T$ provides a local, constant-frequency approximation to $R(\mathbf{x}, t)$.

In terms of spatiotemporal frequency, component velocity may then be expressed in the usual way. At a particular location $\mathbf{x}_0$ and time t_0, the local spatial frequency $\mathbf{k}(\mathbf{x}_0, t_0)$ (which is normal to level curves of constant phase at $\mathbf{x}_0$ in the plane $t = t_0$) gives the normal direction

$$\tilde{\mathbf{n}}(\mathbf{x}_0, t_0) \quad = \quad \frac{\mathbf{k}(\mathbf{x}_0, t_0)}{\| \mathbf{k}(\mathbf{x}_0, t_0) \|} . \tag{6.11}$$

From (6.10) and (6.11), note that $\tilde{\mathbf{n}}(\mathbf{x}_0, t_0)$ is equivalent to $\mathbf{n}(\mathbf{x}_0, t_0)$ in (6.5). The corresponding orientation estimate is then $\tilde{\theta}(\mathbf{x}_0, t_0) = \arg[\mathbf{k}^\perp(\mathbf{x}_0, t_0)]$. Similarly, the 2-d normal speed is given by

$$\tilde{v}_n(\mathbf{x}_0, t_0) \quad = \quad \frac{-\omega(\mathbf{x}_0, t_0)}{\| \mathbf{k}(\mathbf{x}_0, t_0) \|} . \tag{6.12}$$

Again, note that $\tilde{v}_n$ in (6.12) is equivalent to the speed α in (6.6b). Finally, from (6.11) and (6.12), the local phase velocity of $R(\mathbf{x}, t)$ is given by

$$\tilde{\mathbf{v}}_n(\mathbf{x}, t) \quad = \quad \tilde{v}_n(\mathbf{x}, t)\, \tilde{\mathbf{n}}(\mathbf{x}, t) \quad = \quad \frac{-\omega(\mathbf{x}, t)\, \mathbf{k}(\mathbf{x}, t)}{\| \mathbf{k}(\mathbf{x}, t) \|^2} , \tag{6.13}$$

which is a standard expression of velocity in frequency space (e.g. see [Adelson and Bergen, 1985; Fleet and Jepson, 1985; Watson and Ahumada, 1985]). It is also equivalent to $\mathbf{v}_n$ provided by (6.6). Thus, we have shown that the expression of component velocity in terms of level surfaces of constant phase is consistent with the expression of component velocity in Section 3.2 using spatial and temporal frequencies.

---

[2]The expected linearity of $\phi(\mathbf{x}, t)$ is discussed in detail in Chapter 9.

6.4 Relationship to Other Approaches

6.4.1 Gradient-Based Techniques

The phase-based definition of component image velocity has much in common with standard gradient-based approaches that use an initial stage of band-pass filtering (e.g. [Fennema and Thompson, 1979; Enkelmann, 1986; Glazer, 1987]).[3] Yet there are important differences. First is the prefiltering of the image sequence by velocity-tuned filters (see Chapter 4). Second, and more importantly, is the stability of phase compared to the filter output with respect to deviations from image translation. We expect that by tracking phase information we obtain a better approximation to the underlying motion field. Finally, although the robustness of phase information discussed in Section 6.1 concentrated on the use of velocity-tuned filters, the same issues concerning the sensitivity of amplitude information apply to other types of band-pass and low-pass filters, and hence to differential approaches.

6.4.2 Zero-Crossing Tracking

It is also of interest to compare the use of phase information with zero-crossing approaches (e.g. [Buxton and Buxton, 1984; Marr and Ullman, 1981; Waxman et al., 1988; Duncan and Chou, 1988]). To begin, note that zero-crossings are also crossings of constant phase. For example, zero-crossings of the sine-Gabor ($\text{Im}[R(\mathbf{x}, t)]$) output are given by (6.3) when $c = n\pi$, $n \in \mathbb{Z}$.[4] However, Daugman (1987) has argued that zero-crossings are insufficiently rich because there exist signals that have discernible structure, yet produce no zero-crossings after band-pass filtering. Mayhew and Frisby (1981) argue that peaks, in addition to zero-crossings, are required to explain binocular stereopsis. Interestingly, like zeros, crests (peaks) are also surfaces of constant phase, and hence a special case of level phase contours. Phase-based techniques are not restricted to specific values of phase, and therefore make better use of the entire signal. Sub-pixel detection and localization of features, such as zero-crossings, are also unnecessary. As a consequence, we expect the density of velocity measurements to be higher than with zero-crossing techniques. For those who match zero-crossing contours over relatively large distances between frames, note that analogous methods exist for phase information [Jenkin and Jepson, 1988; Sanger, 1988].

---

[3]In some sense, all techniques involve prefiltering in that the sensors impose some degree of low-pass smoothing (spatiotemporal integration), independent of whether further discrete smoothing is performed before, or during, numerical differentiation.

[4]Note, however, that zero-crossings of Gabor filters are not identical to zero-crossings of Laplacian-of-Gaussian output, since they are different band-pass filters. It is beyond the scope of this dissertation to show that they share many of the same qualitative properties.

6.4.3   Phase-Difference Techniques

Finally, although discussed in greater detail in Chapter 11, it is worth noting
here that phase-based techniques have also been recently developed for measuring binocular disparity (e.g. [Jenkin and Jepson, 1988; Sanger, 1988]).
Following Jenkin and Jepson (1988), binocular disparity is defined as the
shift necessary to match the phase values of band-pass filtered versions of
the right and left views. It may be estimated using phase differences between views relative to the local wavelength of the filter output. Because the
bandwidths of the filters used in [Jenkin and Jepson, 1988] were relatively
narrow (near one octave), the local wavelength was assumed to equal the
principal wavelength to which the filters were tuned. A similar approach
has been suggested for measuring velocity [Burt et al., 1989]. Errors in such
phase-difference techniques will arise because of differences between the local
wavelength of response and the peak tuning wavelength of the filter. Errors
will also arise from the implicit form of interpolation (between views). That
is, linear interpolation yields substantial error in signal reconstruction unless
the sampling rate is prohibitively high [Gardenhire, 1964].

In principle, the same phase-difference method could also be applied to
consecutive frames of an image sequence. The phase difference between two
frames can be viewed as an approximation to one component of the phase
gradient, based on linear interpolation. The other component of the phase
gradient would be implicit in the assumption that the local wavelength of
the filter output is determined by the filter tuning.

In earlier motion experiments we used linear interpolation to measure all
components of the gradient, thereby removing some errors due to discrepan-
cies between the response wavelength and the filter tuning. Although some
of the errors caused by poor interpolation were detectable, as they produced
estimates far from the frequencies to which the filters were tuned, there was
a significant decrease in the density of accurate measurements. In general,
with respect to any phase-based technique, errors in velocity measurements
can be expected because of

1) input noise,

2) gross deviations from image translation,

3) quantization noise introduced through signal encoding,

4) inaccuracy due to the form of numerical interpolation/differentiation.

In this monograph component velocity was defined in terms of the phase
gradient. This is an improvement over phase-difference techniques both the-
oretically and in practice, since more accurate forms of interpolation and
measurement follow naturally.

Chapter 7

Experimental Results

The approach introduced in Chapter 6 has been implemented and applied to a variety of image sequences. A 3-d graphics package was used to generate a simple geometric environment in which real images were used to create surface texture. Rendering the scene from a sequence of camera positions under perspective projection produced image sequences complete with perspective distortions including shear, dilation/contraction, and rotation. Image speeds ranged between 0 and 4 pixels/frame. To facilitate comparisons among the results of different motions, we concentrate on the *tree* image shown in Figure 7.1. Results from other surface textures are reported in Appendix D. We also report results from sequences with additive Gaussian noise, transparency, and the *Yosemite* sequence used by Heeger (1988). Results on real inputs are reported in Chapter 8 and Appendix D.

7.1   Implementation

Our approach to measuring component image velocity can be stated as:

1) Apply a collection of velocity-tuned filters.
2) For each filter response:
      Measure component image velocity;
      Remove unreliable measurements.

At present, we use only those Gabor filters that provide a decomposition of a single spatiotemporal pass-band of 0.8 octaves, measured at σ_k. The small bandwidth is important because it reduces sensitivity to mean illumination and low frequencies. Natural images have significant amounts of power at low frequencies [Netravali and Limb, 1980; Langer, 1988], which, if passed by the filters, will cause unwanted aliasing after subsampling, and therefore distortion of local phase. The dc amplitude sensitivity for a Gabor with octave bandwidth β measured at σ_k is $e^{-b^2/2}$ where $b = (2^\beta + 1)/(2^\beta - 1)$;

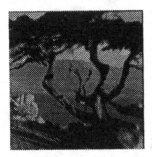

Figure 7.1. **Tree Sequences:** *Frames 10, 20, and 30, from Experiment 5 with camera motion along the line of sight.*

for $\beta = 0.8$ this is roughly 10^{-3}. The residual dc sensitivity can be removed by subtracting a low-pass version of the input (scaled by $e^{-b^2/2}$) from the real (cosine) part of each Gabor output.[1] There were 23 complex kernels in total (see Chapter 4): 6 tuned to speeds about 0 with preferred directions at multiples of 30°; 10 tuned to speeds of $1/\sqrt{3}$ with directions at every 36°; 6 tuned to speeds of $\sqrt{3}$ with directions every 60°; and a *flicker* channel tuned to non-zero temporal frequencies and zero spatial frequency. Because we used spheroidal Gaussian windows, the 46 real 3-d convolutions can be implemented as separable 1-d convolutions. The first and second 1-d stages are common to several of the filters because many of the kernels share the same spatial or temporal frequency tuning. In all, the 46 real 3-d convolutions can be implemented as 75 1-d convolutions [Fleet, 1988]. The organization was designed to be scale invariant in which each frequency band has a similar arrangement of filter tunings with respect to orientation and speed.

Because the set of filters is scale invariant, the velocity resolution available at each scale is constant. The spatiotemporal resolution, however, deteriorates as the spatiotemporal filter support increases. The temporal components of filter support are symmetrical and noncausal, and therefore require a response lag-time equal to the radius of support. In the experiments reported below we used high spatiotemporal frequencies, thereby emphasizing spatiotemporal resolution with a small support width. Unless stated otherwise, the filters were tuned to a spatiotemporal wavelength of 4 pixels (frames). The support radius at one standard deviation was 2.4 pixels (2.4

---

[1]This produces an altered cosine-Gabor kernel of the form $[\cos(\mathbf{x}^T\mathbf{k}_f + t\omega_f) - e^{-b^2/2}]\,G(\mathbf{x}, t; \sigma\mathbf{I})$. The amplitude spectrum of the resulting real and imaginary parts are more alike, and therefore closer to being an exact quadrature pair. With this modified kernel we found a 5% decrease in the magnitude of errors.

frames) in space (time, respectively); the total operator width, out to 3σ, was 15. By comparison, the cones in the human fovea are roughly 20 arc seconds apart and have a temporal integration time of roughly 20 msec. In these terms, the spatiotemporal extent of our filters would be roughly 2 arc minutes and 0.2 seconds. Given this small spatial extent, and the accuracy of the method as demonstrated below, the trade-off between spatiotemporal and velocity resolution does not seem to present a significant limit for practical applications.

The numerical interpolation/differentiation required to measure the phase gradient from the subsampled filter output is discussed in Appendix C. The filter outputs were subsampled with one complex coefficient every $\lfloor\sigma\rfloor$ in space and time. The coefficients were also quantized to 8 bits. This ensures that the representation remains reasonably efficient. The accuracy of velocity estimates reported below would improve if the filter outputs were sampled more densely and not quantized as coarsely.

7.2 Error Measure

All component velocities $v_n\mathbf{n}$ that are generated by the translation of a 2-d textured pattern with velocity $\mathbf{v}$ satisfy

$$\mathbf{v}^T\mathbf{n} - v_n = 0 . \tag{7.1}$$

Equation (7.1) implies that all component velocities consistent with $\mathbf{v}$, represented as $(\mathbf{n}, -v_n)^T$, must lie in the plane normal to $(\mathbf{v}, 1)^T$. Conversely, each component velocity estimate $\tilde{v}_n\tilde{\mathbf{n}}$ constrains the local 2-d velocity to the plane normal to $(\tilde{\mathbf{n}}, -\tilde{v}_n)^T$. Therefore (7.1) forms the basis for the estimation of 2-d velocity $\tilde{\mathbf{v}}$ from estimates of component velocity in Chapter 8. Accordingly, as illustrated in Figure 7.2, an appropriate measure of component velocity error, given a 2-d velocity $\mathbf{v}$ and an estimate of component velocity $\tilde{v}_n\tilde{\mathbf{n}}$, is the angle ψ_ϵ between the estimate and the constraint plane normal to $(\mathbf{v}, 1)^T$. More precisely, the angular error is given by

$$\psi_\epsilon = \arcsin\left(\frac{\mathbf{v}^T\tilde{\mathbf{n}} - \tilde{v}_n}{\sqrt{1 + \|\mathbf{v}\|^2}\sqrt{1 + \tilde{v}_n^2}}\right) . \tag{7.2}$$

Velocity estimates obtained from all 22 filters (excluding the flicker channel) are presented collectively. Two constraints are used to detect *unreliable* velocity estimates:

1. *Frequency Constraint* – The computed local frequency $(\tilde{\mathbf{k}}, \tilde{\omega})^T$, given by the phase gradient must satisfy

$$\| (\mathbf{k}_0, \omega_0)^T - (\tilde{\mathbf{k}}, \tilde{\omega})^T \| < 1.25\,\sigma_k , \tag{7.3}$$

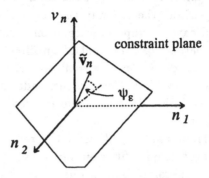

Figure 7.2. **Component Velocity Error:** *In geometric terms, the error ψ_ϵ is the angle between the component velocity estimate and the correct constraint plane determined by the known 2-d velocity* **v**.

where $(k_0, \omega_0)^T$ is the peak tuning frequency, and σ_k denotes the standard deviation of the filter's amplitude spectrum. This means that local frequencies are accepted up to 25% outside the nominal tuning range of the filters.

2. *Amplitude Constraints* – The local signal amplitude must be as large as the average local amplitude, and at least 5% of the largest response amplitude (across all filters at that frame). Local amplitude was computed as a Gaussian weighted average about the pixel in question, averaged over all filters. The standard deviation of the Gaussian was identical to that of the initial Gabor filters. The amplitude constraints detect situations in which there was no significant power at frequencies near $(k_0, \omega_0)^T$, for in such cases noise and quantization error may dominate the local response and make the measurement of the phase gradient very sensitive.

These constraints are important in removing poor velocity measurements, and are discussed further in Chapter 10. In fact, the results in Chapter 10 suggest that a constraint on the derivative of amplitude would be better. However, the experimentation reported in this section was done before the the theoretical results were complete. Rather than rerun all the experiments, we report results from the original algorithm, and note that somewhat better results are obtained with the constraints outlined in Chapter 10.

7.3 Translational Camera Motion

In the first set of experiments we used a class of image velocity fields similar to those considered in [Koenderink and van Doorn, 1976], in which the camera undergoes translational motion with respect to a textured, planar surface.

7.3.1 Camera and Scene Geometry

Assume that the world coordinate system is camera-centred, with the line of sight defined as the Z-axis, a focal length of 1, and a relatively wide field of view subtending 53° (75° diagonally). All images are 150×150 pixels wide. The scene consists of a single planar surface $P(X, Y)$, the gradient of which is given by $(\tan \alpha_1, \tan \alpha_2)^T$, where α_j is specified in degrees. The camera's motion is contained in the XZ-plane, and is expressed as an angle α_c, measured relative to the line of sight; the camera velocity is given by

$$\mathbf{v}_c = v_c \, (\sin \alpha_c, \, 0, \, \cos \alpha_c)^T \, , \qquad (7.4)$$

where v_c is the camera speed expressed in world coordinates (focal length units) per frame. Finally, the distance to the surface along the line of sight is $d(t) = d_0 + v_c t \, (\sin \alpha_c \tan \alpha_1 - \cos \alpha_c)$, where d_0 is the distance at time $t = 0$.

Surface depth, as a function of image location, is given by $Z(x, y) = d(t) \, (1 + x \tan \alpha_1 + y \tan \alpha_2)^{-1}$. From (2.3) it can be shown that the 2-d image velocity *induced* at location $\mathbf{x}$ at time t is

$$\mathbf{v}(\mathbf{x}, \, t) = \frac{v_c(1 + x \tan \alpha_1 + y \tan \alpha_2)}{d(t)} \, (x \cos \alpha_c - \sin \alpha_c, \, y \cos \alpha_c)^T \, . \quad (7.5)$$

From the partial derivatives of (7.5), the magnitudes of divergence $(div \, \mathbf{v})$, curl $(curl \, \mathbf{v})$, and deformation $(def \, \mathbf{v})$ can be determined:

$$div \, \mathbf{v}(\mathbf{x}, \, t) = \frac{\partial v_1}{\partial x} + \frac{\partial v_2}{\partial y}$$

$$= \frac{v_c}{d(t)} \, (\cos \alpha_c (2 + 3x \tan \alpha_1 + 3y \tan \alpha_2) - \sin \alpha_c \tan \alpha_1) \, , \tag{7.6a}$$

$$curl \, \mathbf{v}(\mathbf{x}, \, t) = \frac{\partial v_2}{\partial x} - \frac{\partial v_1}{\partial y}$$

$$= \frac{v_c}{d(t)} \, (\cos \alpha_c (y \tan \alpha_1 - x \tan \alpha_2) + \sin \alpha_c \tan \alpha_2) \, , \tag{7.6b}$$

$$def \, \mathbf{v}(\mathbf{x}, \, t) = \sqrt{\left(\frac{\partial v_1}{\partial x} - \frac{\partial v_2}{\partial y} \right)^2 + \left(\frac{\partial v_2}{\partial x} + \frac{\partial v_1}{\partial y} \right)^2}$$

$$= \left| \frac{v_c}{d(t)} \right| \sqrt{x^2 + y^2 + 1} \, \sqrt{\tan^2 \alpha_1 + \tan^2 \alpha_2} \, . \tag{7.6c}$$

These quantities are of interest in determining the extent to which the projected image velocity deviates from a model of local translation. Note that for $\alpha_c \neq 90$ the velocity field is quadratic. Image speeds can vary significantly throughout the image, as does the direction of motion near the focus of expansion. Also note that these quantities change nonlinearly through time as the distance to the surface $d(t)$ changes.

Two types of translational motion are reported in detail: *1)* with the camera moving perpendicular to the line of sight, as if one were looking at the ground while moving, or out the window of a train ($\alpha_c = 90$); and *2)* with the camera moving along the line of sight ($\alpha_c = 0$). The camera and scene parameters, insofar as they change with each image sequence, are given below.

7.3.2 Side-View Motion

Experiment 1 ($\alpha_c = 90$; $\alpha_1 = \alpha_2 = 0$; $d_0 = 15$; $v_c = 0.075$): The first sequence most closely resembles image translation as the surface is perpendicular to the line of sight and image velocity is constant: $\mathbf{v} = (0.75, 0)^T$ pixels/frame. Figure 7.3 (*top-left*) shows the histogram of the component velocity errors. The inset gives the proportions of the accepted estimates that had errors less than 1°, 2°, and 3° (in absolute value). Figure 7.3 (*top-right*) shows mean error and standard deviation bars as functions of the distance between the estimated local frequencies and the principal filter tunings of the respective channels from which they were obtained (as in (7.3)). Notice the increase in error as the distance from the filter tuning increases. Although a frequency cut-off of $1.25\sigma_k$ was used to select estimates to compute the histogram in Figure 7.3 (*top-left*), it is clear from Figure 7.3 (*top-right*) that the errors are still reasonably well-behaved beyond this boundary. Up to the cut-off most errors are less than 1°.

Figure 7.3 (*middle*) shows the component velocity error behaviour as a function of the estimated orientation. Notice the relatively even distribution of errors as a function of orientation. This is an important property for any scheme used to infer 2-d velocity from the component estimates. Figure 7.3 (*bottom*) shows the distribution of errors across the image. Intensity is proportional to the average (absolute) error per estimate; regions that are entirely black are those containing no estimates that satisfied the frequency and amplitude constraints. Over 70% of the pixels had at least one component velocity estimate. In this, as well as in the following experiments with the tree image, there were over 10^4 estimates accepted; it was often the case that 3 or more estimates occurred at a single pixel.

Experiment 2 ($\alpha_c = 90$; $\alpha_1 = 15$; $\alpha_2 = 0$; $d_0 = 13$; $v_c = 0.173$): The second sequence involved faster speeds and a non-zero surface gradient. This

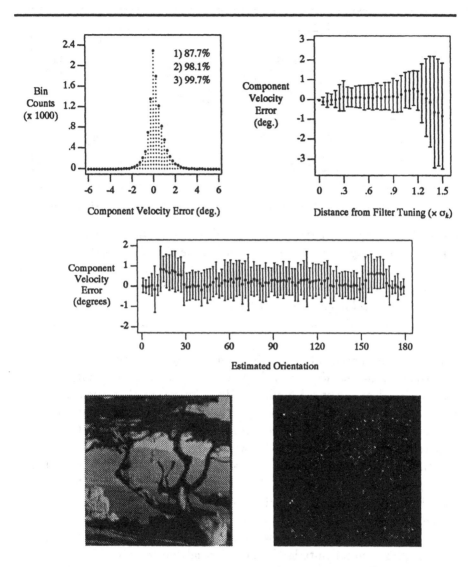

Figure 7.3. **Experiment 1** (side-view motion): *(top-left) Histogram of velocity errors (7.2), with the proportions of estimates with errors less than 1°, 2° and 3°. (top-right) Mean error and standard deviation bars as a function of distance between estimated local frequencies and corresponding filter tunings (relative to the width of the amplitude spectra, σ_k). (middle) Mean velocity error and standard deviation bars as a function of estimated orientation. (bottom-left) The frame at which velocity was computed. (bottom-right) Average absolute error per estimate (as intensity) as a function of image location. Black regions contain no estimates.*

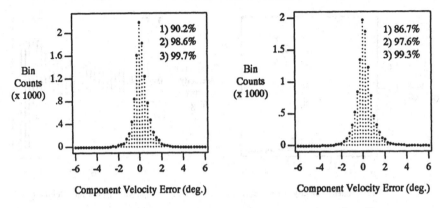

Figure 7.4. **Experiments 2 and 3** (side-view motions): *Histograms of component velocity errors for Experiments 2 (left) and 3 (right). As above, the inset shows the proportions of estimates with errors less than 1°, 2° and 3°. In both cases the error statistics are again a function of distance between the estimated local frequencies and the filter tunings.*

produces a speed gradient in the direction of image velocity, and differs from Experiment 1 in that $div\,\mathbf{v}$ and $def\,\mathbf{v}$ are non-zero (but constant). As $|\alpha_1|$ increases so does the speed gradient. Here, image speeds ranged from 1.73 pixels/frame on the left side of the image to 2.63 on the right. Despite the faster, non-uniform image velocities, the results are similar to those in Experiment 1. In particular, as shown in Figure 7.4 (*left*), the proportions of estimates with errors below 1°, 2°, and 3° were 90.2%, 98.6% and 99.7%.

Experiment 3 ($\alpha_c = 90$; $\alpha_1 = 0$; $\alpha_2 = 20$; $d_0 = 13.5$; $v_c = 0.135$): The final side-view sequence had a vertical surface gradient. This produces a speed gradient perpendicular to the direction of image velocity. As a consequence, $curl\,\mathbf{v}$ and $def\,\mathbf{v}$ were non-zero while $div\,\mathbf{v} = 0$. The perceived effect is motion parallax, the magnitude of which depends on $|\alpha_2|$; in this case it was quite visible. The image speeds ranged from 1.2 pixels/frame at the top of the image to 1.8 at the bottom. Figure 7.4 (*right*) shows that the results are again similar to those above, with 86.7%, 97.6%, and 99.3% of the estimates having errors less than 1°, 2°, and 3°. Results on other image sequences with more curl and shear are shown in Appendix D.

7.3.3 Front-View Motion

In the next two experiments the camera moved along the line of sight. Image velocities point radially out from the centre of the image (the focus of expansion), with speeds increasing toward the image boundaries; $div\,\mathbf{v}$ is non-zero and varies as a function of time. As a result of the dilation there are local variations in scale, speed, and the direction of image velocity, especially near the focus of expansion. Relative to the accuracy with which we hope to measure velocity, these local changes in the direction of motion, speed, and scale constitute significant deviations from a model of image translation. Also note that there did exist significant structure at spatiotemporal frequencies higher than those to which the filters were tuned. Although relatively high, the frequency range to which the filters were tuned was not at the folding frequency. Furthermore, as time progressed and the camera moved closer to the surface, new structure appeared because the initial frames were (in effect) down-sampled versions of the original. Subsequent frames were rendered by reprojecting the surface (and the texture) at each frame, and not by simply interpolating the first frame.

Experiment 4 ($\alpha_c = \alpha_1 = \alpha_2 = 0$; $d_0 = 16$; $v_c = 0.15$): In the first sequence the surface was perpendicular to the line of sight and the velocity was relatively slow; time to collision was 105 frames. The induced velocity field was spatially isotropic with image speeds up to 1 pixel/frame in the corners. Figure 7.5 shows the results. The measurement density and error behaviour are similar to the first three experiments. In fact, the behaviour of errors as a function of the distance from the principal filter tunings in Figure 7.5 (*top-right*) appears even more consistent, as does the error as a function of estimated orientation. In comparison to the side-view motion sequences above, the velocity field here is dominated by dilation, and the normal velocity estimates are not as accurate to within 1° of the true velocity. However, the proportions of estimates with error less than 2° and 3° are similar.

Experiment 5 ($\alpha_c = 0$; $\alpha_1 = 20$; $\alpha_2 = 0$; $d_0 = 13$; $v_c = 0.2$): The time to collision here was 65 frames and the induced image speeds ranged from 0 in the centre of the image, to 1.33 pixels/frame on the left and 1.9 on the right. With 2-d velocity expressed as a direction vector in space-time (1.11), and speed expressed in degrees (i.e. arctan $\|\,\mathbf{v}\,\|$), this local speed variation amounts to speed differences of close to 1° between neighbouring pixels (about 15° over the entire operator width). In addition, over the width of temporal support, the distance to the surface $d(t)$ decreased by about 20%. As a consequence, $div\,\mathbf{v}$ changes significantly. Figure 7.6 shows the histogram of component velocity errors (*left*) as well as the error behaviour as a function of the distance between local frequency and the peak tuning frequencies of the filters. As above, the errors are still well-behaved. Although the proportion of estimates accurate to within 1° of the true velocity is lower, the proportions of estimates with errors less than 2° and 3° are high (similar to Experiments 1-3). This accuracy is good considering the speed, direction, and scale changes

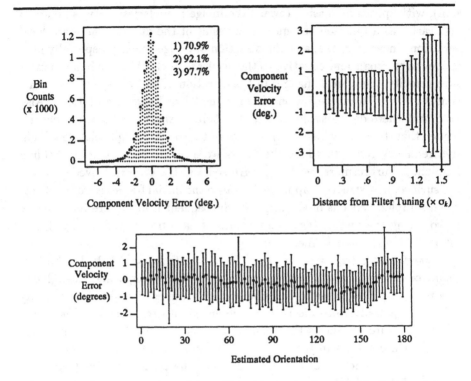

Figure 7.5. Experiment 4 (front-view motion): *Image velocities point radially out from the line of sight. Speed increases from 0 in the centre to 1 pixel/frame at the corners. (top left) Histogram of errors with inset showing proportions of errors less than 1°, 2° and 3°. (top right) Mean error and standard deviation bars as a function of distance between the estimated local frequencies and the filter tunings. (bottom) Mean component velocity error and standard deviation bars as a function of the estimated orientation.*

within the spatiotemporal support width of the filters. The distribution of errors over the image is again similar to that shown in Figure 7.3.

7.4   Image Rotation

The image velocity fields considered in the first five experiments were dominated by translation and dilation (despite the non-zero curl in Experiment 3). The next experiment deals explicitly with image rotation.

Experiment 6 (counter-clockwise rotation, 1°/frame): The image velocity fields that result from instantaneous camera rotation (with no translational component) do not depend on the depth of scene points [Longuet-Higgins

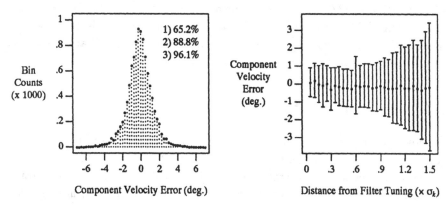

Figure 7.6. **Experiment 5** (front-view motion): *Image velocity is 0 at the centre, 1.33 pixels/frame on the left, and 1.9 on the right. (left) Histogram of errors with inset showing the proportions of errors less than 1°, 2° and 3°. (right) Mean error and standard deviation bars as a function of distance between the estimated local frequencies and the filter tunings.*

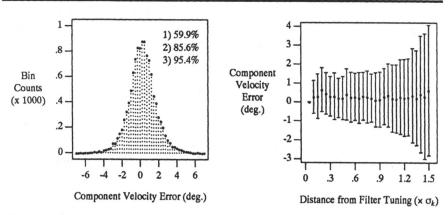

Figure 7.7. **Experiment 6** (image rotation): *Image speeds ranged from 0 at the centre to 1.31 at the edges of the image (1.85 in the corners). (left) Histogram of component velocity errors. (right) Error behaviour as a function of distance between estimated local frequencies and the filter tunings.*

and Prazdny, 1980]. Therefore, we tested only the simplest case in which the camera rotates while the planar surface remains normal to the line of sight. With rotation of 1°/frame, and an image size of 150 × 150 pixels, the image speeds ranged from 0 in the centre of the image to 1.31 pixels/frame at the edges (1.85 in the corners). The results shown in Figure 7.7 are similar to the dilation sequence in Experiment 5.

7.5   Additive Noise

We now consider the robustness of phase information when additive noise degrades the input. Spatiotemporal white Gaussian noise was added to several image sequences to demonstrate the error in velocity estimates as a function of the noise level. Here we report results from two sequences, namely, Experiments 2 and 5. The noise was mean zero with standard deviations σ_n up to 50. The original images had 8 bits/pixel (see Figure 7.8 (*top*)).

Figure 7.8 (*middle and bottom*) shows the decrease in the proportions of errors falling within 1°, 2°, and 3° of the correct velocity as a function of σ_n. As expected, the accuracy deteriorates with increased noise levels. However, the total number of estimates that survived the frequency and amplitude constraints remained roughly constant, and the deterioration occurred smoothly and relatively slowly. The high proportion of estimates within 3° of the true velocity is especially encouraging. The sharper increase in errors in Figure 7.8 (*bottom*) arises from the generally poorer performance in Experiment 5, in conjunction with the tuning of the filter to the low contrast horizontal image structure near the centre of the image. As is clear in Figure 7.8 (*top*) these regions are easily degraded by small amounts of noise.

7.6   Rotating Sphere and Yosemite Sequence

The next two experiments involve synthetic image sequences depicting more complex scene structure. The first sequence contains a rotating, textured sphere (Figure 7.9). The second is the *Yosemite* image sequence used by Heeger (1988) (Figure 7.10).

For the rotating sphere the image size was 200 × 200, and the angular field of view was 40° (54° diagonally). The distance between the centroid of the sphere and the focal point of the camera, was 4 times the radius of the sphere. The rotation was 1.5°/frame about its centroid, with the axis of rotation given by $(45°, 35°)^T$ in standard spherical coordinates. This induces image speeds of up to 2.6 pixels/frame along the equator, and 0 at the fixed point (see Figure 8.5). Along the boundary of the sphere there is a large amount of noise caused by the the texture-mapping algorithm. In addition,

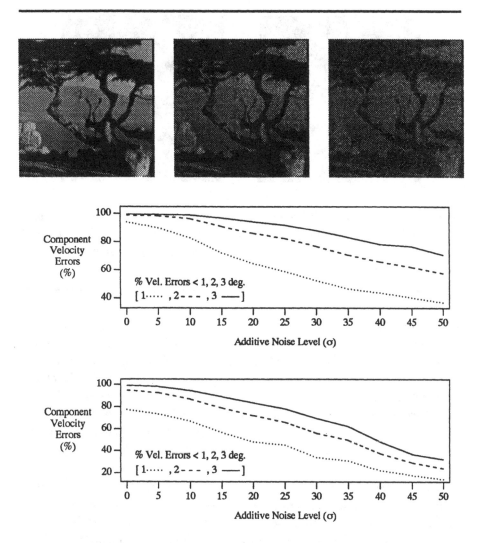

Figure 7.8. **Effects of Additive Gaussian Noise:** *(top) An original image, and the same image with additive mean-zero white Gaussian noise with standard deviations $\sigma_n = 15$ and $\sigma_n = 40$. The bottom two plots show the proportions of estimates with errors below $1°$, $2°$ and $3°$ as a function of σ_n for single filters from the family. (middle) Image sequence of Experiment 2, and a filter tuned to component velocities about 1.732 pixels/frame down to the right. (bottom) Image sequence of Experiment 5, and a filter tuned to component velocities about 0.577 pixels/frame up toward the top.*

Figure 7.9. **Rotating Sphere:** *2-d image velocities were up to 2.6 pixels/frame along the equator and zero at the fixed point. (left) One frame of the image sequence. (right) Average component velocity error as a function of image location.*

because of the loss of resolution caused by the filter support, about 9% of the estimates lie outside the projected boundary of the sphere: the extreme errors occur in this region. Within the boundary of the sphere the accuracy is similar to that in Experiments 5 and 6, with 58%, 80%, and 89% of the estimates having errors of less than 1°, 2° and 3°. Outside the boundary of the sphere, the estimates are still consistent with the motion of the sphere as can be seen from the estimated 2-d velocity in Chapter 8. When real images were texture-mapped onto the sphere, instead of the checker-board pattern, the results were similar.

The *Yosemite* sequence contained only 15 frames. Therefore a smaller spatiotemporal window of support (7 pixels-frames) was used with a corresponding decrease in the spatiotemporal wavelength to which the filters were tuned. All other parameter settings and thresholds were identical to those used above. The results are shown in Figure 7.10. Performance with the *Yosemite* sequence is worse than Experiments 5 and 6. Like Heeger's results, and those with the rotating sphere, most of the extreme errors lie on the occlusion boundaries. A few errors are due to the cloud movement for which we had no exact velocity information. However, note that most of the sky region is dominated by relatively low spatiotemporal frequencies to which the filters were relatively insensitive; in a more complete implementation (with more than one scale) the cloud motion would be detected. Errors were also due to aliasing in the lower left where velocities are greater than 3 pixels/frame, and numerical error as the filters were tuned to the highest end of the frequency

Figure 7.10. **Yosemite Sequence:** *Velocities in the Yosemite sequence were predominately toward the left with speeds ranging from 0 to 4. The clouds moved (non-rigidly) to the right at 1 pixel/frame. (left) One frame from the sequence. (right) Average (absolute) component velocity error as a function of image location.*

spectrum. If we exclude the sky region, we find that about 85% of the image has at least one component velocity estimate, and that 60%, 79%, and 87% of the estimates have errors less than 1°, 2° and 3°.

7.7 Transparency

Our final experiment in this section concerns the issue of velocity resolution when multiple motions exist in local neighbourhoods. Two samples of white Gaussian noise (mean zero with $\sigma = 250$) were combined additively. The first covered the entire image and was stationary. The second, masked by the characteristic function shown in Figure 7.11 (*top-left*), moved with speed 1.5 pixels/frame and direction 30°. This roughly simulates the motion of a textured object viewed through a window on which there is the reflection of a stationary textured surface. It does not simulate the complex behaviour of photon transmission through nonlinear refractive materials. Our goal is simply to show that reliable estimates of component velocity can be obtained within a spatial region in which more than one motion exists. Conversely, note that any technique based on purely spatial image properties (e.g. zero-crossings of $\nabla^2 G$) will yield incorrect results.

To demonstrate velocity resolution, the component velocity estimates were divided into three groups according to whether they were consistent

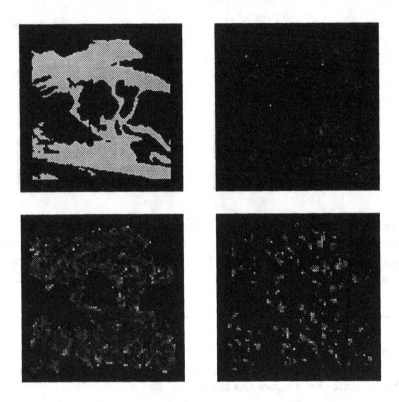

Figure 7.11. **Multiple Velocities:** *(top-left) The characteristic function for the moving object. The other three images show average (absolute) component velocity error as a function of image location for the three groups of estimates that are consistent with (top-right) the stationary window, (bottom-left) the moving object, and (bottom-right) neither.*

with the stationary window, the moving object, or neither. Consistency was defined in terms of the error between the component velocity estimates and the constraint planes of the 2-d velocities (7.2); each estimate was grouped with the constraint plane to which it was closest. If it was not within 10° of either it was deemed inconsistent. Estimates with spatial orientations between 25° and 35° were omitted as they are consistent with both 2-d velocities.

Figure 7.11 shows the average (absolute) error ($|\psi_c|$ in (7.2)) as a function of image location for the three groups of estimates. As above, pixels are black where there are no estimates. In comparing the three images note first that the estimates consistent with the window cover the entire image, while those consistent with the moving object coincide essentially with the characteristic

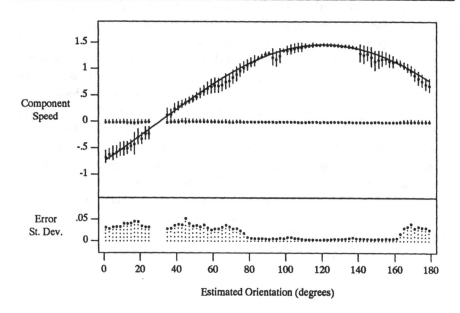

Figure 7.12 **Velocity Resolution:** *(top) Average component speed and standard deviation bars are shown, as a function of the estimated orientation, for the estimates that are consistent with the window and the object, as in Figure 7.11 (top-right and bottom-left). The continuous curve shows the correct speed for the object. The window was stationary. (bottom) The standard deviation of the errors for the window is magnified here to show the improved accuracy that occurs when the two signals differ more in speed.*

function. The mean (absolute) errors per pixel (averaged over pixels with at least one estimate) for those estimates consistent with the window and the object are 0.75° and 1.9° respectively, with standard deviations 0.7 and 1.6. By way of comparison, the measurements from the window are as accurate as those in Experiments 1–4, while those of the object are about as accurate as Experiments 5 and 6. Errors are generally larger where the window and object overlap. Image locations at which estimates are inconsistent with both surfaces are sparse. The mean number of velocity estimates per pixel (averaged over pixels with at least one estimate) for the three cases were 3.9 ($\sigma = 1.1$) for the stationary window; 2.8 ($\sigma = 1.4$) for the moving tree; and 1.1 ($\sigma = 0.3$) for the inconsistent estimates. Therefore, it is rare to find more than one inconsistent estimate at a given location – that is, poor estimates are sparse.

Finally, Figure 7.12 illustrates the same results in another way to emphasize the velocity resolution obtained with the different filters. Figure 7.12 *(top)* shows the average component speed (with standard deviation bars) for the two groups of velocity estimates that were consistent with the object and the window. The removal of velocity estimates with orientations between 25° and 35° is evident. The solid line shows the correct component speed as a function of orientation for the object. The window was stationary and therefore has a speed of zero at all orientations. Figure 7.12 (bottom) shows a magnified view of the standard deviation of the speed estimates for the window. Notice that, as the speeds of the window and the object diverge, the accuracy of the measurements, and our ability to exploit the velocity resolution, increases. Thus, although the experiments with single motions (Figures 7.3 and 7.5) showed the errors to be spread nearly evenly across orientations, with multiple velocities we expect greater measurement error when the velocities are similar.

Chapter 8

Computing 2-D Velocity

Another way to demonstrate the accuracy of the component velocity measurements is to show the accuracy with which estimates of 2-d velocity could be computed by a subsequent stage of processing. A linear model for $\mathbf{v}(\mathbf{x}, t)$ in each local neighbourhood was determined from collections of component velocity estimates in the neighbourhood. As discussed in Chapter 5, this approach is limited because it presupposes that the local component velocities reflect the motion of a single smooth surface in the scene; the model fails whenever there are multiple velocities, or when there are significant outliers, to which least-squares minimization is sensitive. In essence, the approach taken here tacitly assumes that some form of preliminary segmentation of the component velocity estimates has already occurred.

The segmentation of component velocity estimates requires further research, which is beyond the scope of this monograph. Our primary purpose here in computing 2-d velocity is to demonstrate the accuracy and robustness of the component velocity estimates. Moreover, it helps us to evaluate the results of the technique when applied to real image sequences for which the true 2-d velocity fields are unavailable.

8.1 Basic Least-Squares Technique

The derivation of the approach follows from (7.1), a linear constraint on component velocity and 2-d velocity. In what follows we assume that the available collection of component velocity estimates in the neighbourhood of a point $\mathbf{x}_0$ at time t reflects the relative motion of a smooth surface (as in (2.4)), and that the local velocity field may be approximated by

$$\tilde{\mathbf{v}}(\mathbf{x}, t) = (a_0, b_0)^T + \begin{pmatrix} a_1 & a_2 \\ b_1 & b_2 \end{pmatrix} \tilde{\mathbf{x}}, \qquad (8.1)$$

where $\tilde{\mathbf{x}} = \mathbf{x} - \mathbf{x}_0$. A linear model of the local velocity field is not used as commonly as a model of constant velocity (but see [Burt et al., 1989; Waxman and Wohn, 1985; Adiv, 1985]).

With respect to the neighbourhood of $\mathbf{x}_0$ at time t_0, an estimate of component velocity $\tilde{v}_n \tilde{\mathbf{n}}$ at location $\mathbf{x}$ yields the linear constraint

$$(\tilde{n}_1, \ \tilde{n}_1 \tilde{x}, \ \tilde{n}_1 \tilde{y}, \ \tilde{n}_2, \ \tilde{n}_2 \tilde{x}, \ \tilde{n}_2 \tilde{y}) \ \mathbf{a} \ = \ \tilde{v}_n \ , \qquad (8.2)$$

where $\mathbf{a} = (a_0, \ a_1, \ a_2, \ b_0, \ b_1, \ b_2)^T$ denotes the vector of unknowns, $\tilde{x} = (x - x_0)$, and $\tilde{y} = (y - y_0)$. These constraints provide a system of linear equations $N\mathbf{a} = \mathbf{s}$ where $\mathbf{s}$ denotes the vector of normal speeds $\tilde{v}_n$. Given more than six local measurements, the solution to the over-determined system, $\tilde{\mathbf{a}}$, minimizes (in a least-squares sense) $\| N\mathbf{a} - \mathbf{s} \|^2$.

8.2   Provisional Implementation

The estimates of component velocity that satisfied the constraints in Section 7.2 were collected about each pixel (on the subsampled grid) within a radius of 2 pixels.[1] A singular-value decomposition (SVD) was used to determine the conditioning of the resultant system. Large condition numbers[2] indicate the possibility of significant error amplification because the sensitivity of least-squares solution to errors in the data is proportional to the κ. In our case, high condition numbers reflect an insufficient amount of local structure from which $\mathbf{v}$ could be computed (e.g. because of the aperture problem); here we used a threshold of $\kappa > 10$ to indicate unreliable velocity estimates. From a geometric perspective, the restriction on κ means that a minimal distribution of local orientations is necessary for the computation of $\mathbf{v}$ (8.1). For example, in just two dimensions $\kappa \leq 10$ means that the local component velocities must cover at least 5 degrees of the constraint plane depicted in Figure 7.2. In the experiments reported below κ was often between 5 and 10; in most cases $\kappa \leq 10$ produced a dense set of 2-d velocity estimates. When there was sufficient local information, the least-squares system was solved using the pseudo-inverse provided by the SVD. Finally, estimates of $\mathbf{v}$ were discarded when the residual error $\| N\tilde{\mathbf{a}} - \mathbf{s} \| / \| \mathbf{s} \|$ was greater than 0.5. The accuracy of the estimated translational velocities $(a_0, b_0)^T$ are reported

---

[1]When a much smaller neighbourhood was used to collect component velocities the system was often ill-conditioned because of the aperture problem. This remained true even for a model of constant velocity. Also note that a model of constant velocity sometimes produces poorer estimates of $\mathbf{v}$ because the spatial dependence of the component velocity estimates is disregarded (unless they are uniformly distributed about the centre of the neighbourhood).

[2]The condition number κ of a matrix is the ratio of the largest singular-value to the smallest.

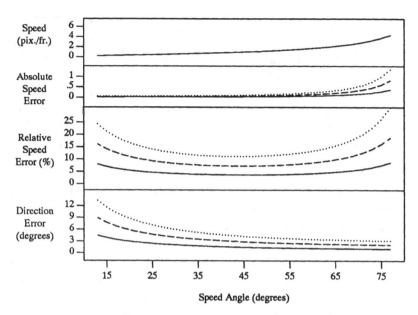

Figure 8.1. **Speed in Degrees vs. Pixels/Frame:** *For fixed angular velocity errors ψ_ϵ in (8.3), errors in pixels/frame depend on angular speed. With* **v** *represented as a unit direction vector in space-time, we can view velocity in spherical coordinates as in (1.12), in terms of angular speed θ_v and direction θ_x. From top to bottom, with $\psi_\epsilon = 1°$ (solid), $2°$ (dashed), and $3°$ (dotted), the four panels correspond to:*

a) *Speed in pixels/frame:* $\tan(\theta_v)$;
b) *Absolute speed errors (pixels/frame):* $\tan(\theta_v) - \tan(\theta_v + \psi_\epsilon)$;
c) *Relative speed errors:* $100.0(\tan(\theta_v) - \tan(\theta_v + \psi_\epsilon))/\tan(\theta_v)$;
d) *Maximum error in direction of motion (in degrees):* $\psi_\epsilon/\sin(\theta_v)$.

---

below. Estimates of ∇v were accurate (with errors of 10%-15%) only for larger neighbourhoods with radii of 4 or 5 pixels.

The error between the known 2-d velocity **v** and the computed translational velocity $\tilde{\mathbf{v}}$ was taken to be the angle between their space-time direction vectors $(\mathbf{v}, 1)^T$ and $(\tilde{\mathbf{v}}, 1)^T$:

$$\psi_\epsilon = \arccos\left(\frac{\mathbf{v}^T\tilde{\mathbf{v}} + 1}{\sqrt{1+ \parallel \mathbf{v} \parallel^2}\sqrt{1+ \parallel \tilde{\mathbf{v}} \parallel^2}}\right). \tag{8.3}$$

Although this measure of error is somewhat non-standard, it is consistent with that used for component velocities (7.2), and it complements the notion

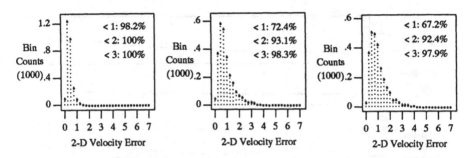

Figure 8.2. **Histograms of 2-d Velocity Errors:** *From left to right are the histograms of 2-d velocity errors for Experiments 1, 5 and 6, the component velocity errors of which are shown in Figures 7.3, 7.6 and 7.7.*

of velocity as space-time orientation. In particular, remember that component velocity errors are typically concentrated about the constraint plane (7.1), and are uniformly distributed over orientation (Figures 7.3 and 7.5). Therefore, we expect that for reasonably well-conditioned systems, the estimated 2-d velocities will be concentrated within a cone about the true velocity $(\mathbf{v}, 1)^T$ in space-time; the opening angle of the cone depends on the magnitude of errors in component velocity estimates and the distribution of component velocity estimates over the constraint plane. Finally, note that, for a fixed angular error (8.3), absolute velocity error $\| \mathbf{v} - \tilde{\mathbf{v}} \|$ and relative error $\| \mathbf{v} - \tilde{\mathbf{v}} \| / \| \mathbf{v} \|$ depend on speed $\| \mathbf{v} \|$. These relationships are illustated in Figure 8.1.

8.3   Experimental Results

8.3.1   Experiments 1, 5 and 6

Figure 8.2 shows histograms of 2-d velocity errors for the 2-d velocities computed from Experiments 1, 5, and 6. Experiments 1–3, with predominantly translational velocity fields, produced the most accurate estimates of component velocity, and hence the most accurate 2-d velocity estimates. Figure 8.2 (*left*) is also characteristic of the results of Experiments 2 and 3. The cases of dilation and rotation (Experiments 5 and 6) were not handled as well as the first four experiments. Despite this, the errors are almost all less than two degrees. Figure 8.3 (*top*) shows the estimated 2-d velocity fields for Experiments 5 and 6. The estimated 2-d velocities in both cases are sufficiently accurate that the vector differences between the true and the estimated velocities are not resolvable at this scale. Therefore the true velocities

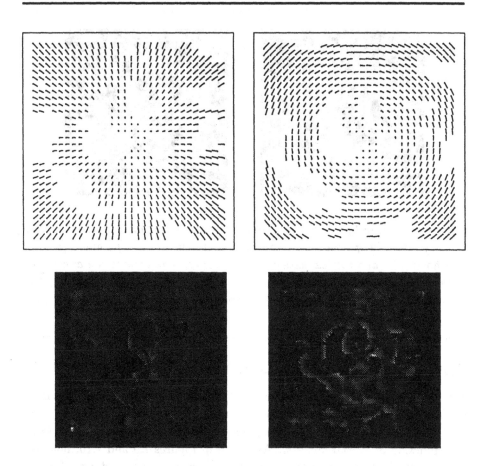

Figure 8.3. **Velocity Results for Experiments 5 and 6:** *(top) Computed 2-d velocity fields from Experiments 5 and 6. (bottom) 2-d velocity errors for Experiments 5 and 6 are shown as a function of spatial location.*

and vector differences are not shown. Figure 8.3 (*bottom*) shows the absolute angular velocity error (8.3) as a function of spatial location. The errors are concentrated along the boundaries of regions without measurements where the component velocities are not as accurate, and where the least-squares systems were less well-conditioned. In particular, note that the errors near the flow singularities are not significantly larger than in other areas. Finally, Figure 8.4 shows histograms of angular differences between the true and esti-mated velocities in spherical coordinates for Experiments 5 and 6. This helps to show that the errors are distributed evenly about the true velocities, and therefore the angular measure of error (8.3) is appropriate.

Figure 8.4. **Velocity Errors for Experiments 5 and 6:** *Histograms of 2-d velocity errors in spherical coordinates for Experiments 5 (left) and 6 (right). The horizontal and vertical axes correspond to error in orientation and speed. If the true and estimated 2-d velocities are given by (θ_x, θ_v), and $(\tilde{\theta}_x, \tilde{\theta}_v)$, then we increment locations $((\theta_x - \tilde{\theta}_x)\sin\theta_v, \theta_v - \tilde{\theta}_v)$. Both histograms have widths and heights of 4°.*

8.3.2   Rotating Sphere and Yosemite Sequence

We also computed 2-d velocity fields for the rotating sphere and the Yosemite sequences. Figures 8.5 and 8.6 show the estimated velocities, the true velocities, the vector differences between them, and the 2-d velocity errors (8.3) as a function of spatial location. As shown in Figures 7.9 and 7.10, both sequences produce estimates that spread across occlusion boundaries. This is evident in Figures 8.5 and 8.6 as well. However, these estimates are generally consistent with the nearby surface motion.

Taking into account only those estimates within the boundary of the sphere, the proportions of 2-d estimates with errors less than 1°, 2°, and 3° were 72%, 88%, and 93%. Over this region the errors are generally uniform, and the flow singularity is handled well. Also notice the vertical regions containing no 2-d velocity estimates, where, from Figure 7.9 (*right*), it is clear that there were component estimates. Interestingly, these regions fall precisely down the centres of the wide rectangular checkers (cf. Figure 7.9 (*left*)). Relative to the tuning of the filters to high spatiotemporal frequencies, and the small radius within which the component velocity estimates were combined to estimate 2-d velocity, the structure in these regions is essentially one dimensional. About the equator, however, the checkers are somewhat foreshortened and the image speeds are faster so that lower spatial frequencies will stimulate those filters tuned to faster speeds. Hence there is a greater

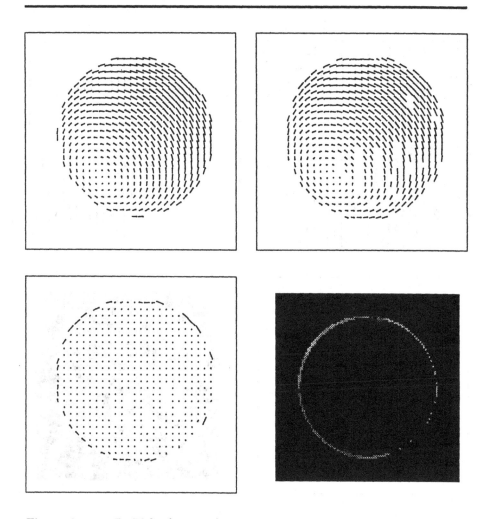

Figure 8.5. **2-D Velocity Estimates for Rotating Sphere:** *(top-left) True (induced) 2-d velocity field from the Rotating Sphere Experiment. One image from the sequence is shown in Figure 7.9. (top-right) Estimated 2-d velocities. (bottom-left) The vector difference between the two fields. (bottom-right) 2-D velocity error as a function of image location.*

density of significant filter activity and sufficient structure for the estimation of 2-d velocity. The regions currently without 2-d estimates would be filled in by filters tuned to lower spatiotemporal frequencies in a more complete implementation with more than one scale.

For the Yosemite sequence, if we omit errors just above the horizon in the

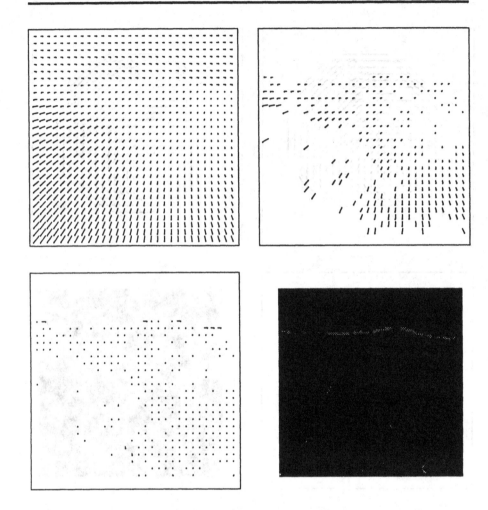

Figure 8.6. 2-D Velocity Estimates for Yosemite Sequence: *(top-left) True (induced) 2-d velocity field from the Yosemite Sequence, one frame of which is shown in Figure 7.10. (top-right) Estimated 2-d velocities. (bottom-left) The vector difference between the two fields. (bottom-right) 2-D velocity error as a function of image location.*

sky region, the proportions of 2-d velocity estimates with errors less than 1°, 2°, and 3° are 45%, 71%, and 82%. Although these results are not as good as those above, many of the poor 2-d velocity estimates are caused by poor conditioning or a poor least-squares fit. For example, when we discard all estimates with condition numbers greater than 5 (instead of 10), and residual errors greater than 0.1 (instead of 0.5), the proportions of errors below 1°, 2°, and 3° increase to 63%, 89% and 95% while the total number of estimates drops by about one third. This is important as it means that the errors in the 2-d velocity estimates are due mainly to the conditioning of the least-squares system, and not to inaccuracy in the component velocity estimates (which is our principal concern). Finally, these results compare favourably with those obtained with Heeger's model, for which the histogram of errors was relatively flat with about 90% of the estimates having errors less than 25°. The proportions of estimates with errors less than 5°, 10°, and 15° were only approximately 30%, 60%, and 80%. However, Heeger used different spatiotemporal scales and larger spatial support which led to different results. In particular, his technique gives results in areas (e.g. the sky and the lower left) where the present method does not.

8.3.3 Real Image Sequences

Finally, we show velocity estimates obtained from real image sequences. Unfortunately, accurate information about the 2-d motion fields for these sequences is unavailable, which rules out quantitative performance analysis. However, real image sequences remain the standard by which we compare; they serve to uncover problems not encountered with synthetic inputs. Here we report results obtained from the *Hamburg Taxi Sequence*, which has been cited extensively in the literature [Nagel, 1983; Enkelmann, 1986; Nagel and Enkelmann, 1986]. Results from other sequences are given in Appendix D. In all cases we used the same filters and parameters that were used in all but the *Yosemite* sequence above.

There are 46 frames in the Taxi sequence. Velocity was computed at frame 21, shown in Figure 8.7 (*left*). There are four moving objects in the scene: the taxi which produces image speeds of just under 1 pixel/frame; the Golf in the lower left which produces speeds of about 3.75 pixels/frame; the van in the lower right, which is partially occluded and exhibits speeds similar to the Golf; and a pedestrian in the upper left who moves down to the left at about 0.3 pixels/frame. Branches of two trees are also moving slowly.

Figure 8.7 (*right*) shows where (in the image) estimates of component velocities were obtained. Black areas denote regions within which no estimates occur. The darker grey areas denote regions in which all estimates had normals speeds between 0 and 0.15 pixels/frame. The brighter areas show regions in which there existed estimates with normal speeds greater than 0.15 pixels/frame. This crude quantization helps to illustrate the distribution of component velocity estimates and the regions of higher speeds that correspond to the four primary moving objects.

Figures 8.8 and 8.9 show the 2-d velocities that were computed from the

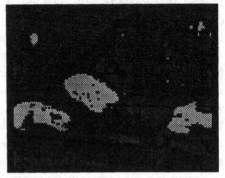

Figure 8.7. **Hamburg Taxi-Cab Sequence:** *(left) Frame 21 from the sequence. (right) Regions with no component velocity estimates are black. Grey and white regions correspond to regions containing component velocity estimates with normal speeds between 0 and 0.15 pixels/frame and greater than 0.15 pixels/frame respectively.*

component estimates. Figure 8.8 (*top*) shows the estimated 2-d speed (shown as intensity) as a function of image location. The vector fields corresponding to the two boxed areas are then shown below (blown up so that the individual vectors are resolvable). Similarly Figure 8.9 shows the velocity fields for the Golf and the van. The black dots (not joined to vectors) represent speeds close to 0. Not all regions with component velocity measurements yielded 2-d estimates because of the local nature of the computation. This is particularly evident along the rear windows of the taxi cab. Also, from Figure 8.7 notice the large number of estimates in low contrast regions (e.g. the street marking to the right of the taxi). The robustness of local phase behaviour as compared to amplitude is especially clear in areas of low contrast.

Finally, it should be emphasized that no smoothing has been applied to these measurements, other than that implicit in the least-squares minimization. This is important in comparing the results to other techniques that use smoothness constraints to propagate the influence of the raw measurements throughout the image (e.g. [Nagel and Enkelmann, 1986]). Similarly, it is important to remember that our objective in computing 2-d velocity from the component velocity estimates was to illustrate the accuracy of the component velocity estimates. As mentioned in the introduction, the integration of local measurements implicitly assumes that all the component velocity estimates arise from the same physical object. Here, a unique 2-d velocity arising from a single object is assumed within each local neighbourhood. The velocity es-

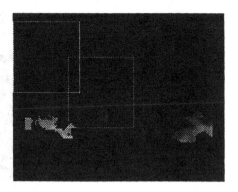

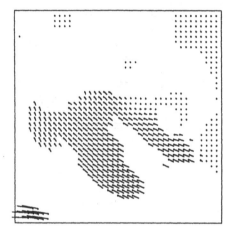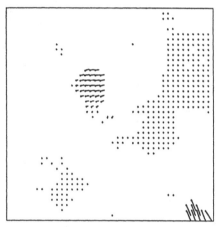

Figure 8.8. **Taxi Sequence Flow Fields:** *(top) Speed is shown as a function of image location; regions without estimates of* v *are black. (bottom) Vector fields for the regions marked by the white boxes in the top image. Each region has been enlarged and scaled for better visibility; the velocity estimates about the taxi and the pedestrian had speeds scaled to 0 – 1.0 pixels/frame, and 0 – 0.3 pixels/frame.*

timates near the front of the van in Figure 8.9 (*bottom-right*) show that this is not always appropriate. In this case, measurements from a tree branch and the van are combined, and yield a good fit 2-d velocity (with low residual error). Subsequent smoothing of the velocity estimates would aggravate the problem. A framework for integrating velocity measurements in the face of multiple moving objects is an important area for further research.

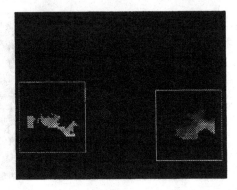

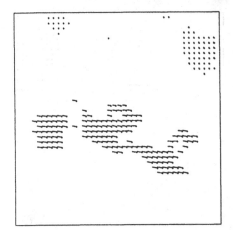
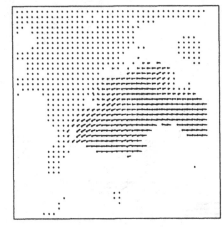

Figure 8.9. **Taxi Sequence Flow Fields:** *(top) Speed is shown as a function of image location. The white boxes delineate the regions for which the 2-d velocities are enlarged as vector fields. (bottom) Velocity estimates about the Golf and the van with speeds scaled to 0 – 3.5 pixels/frame in both cases.*

8.4   Discussion

In Chapter 6, component image velocity was defined in terms of the gradient of the phase of the output of individual velocity-tuned linear filters. The computation of component velocity involved two main stages of processing:

- The time-varying image sequence is first represented with a family of constant-phase velocity-tuned filters.

- The local phase gradient is then measured from the output of the individual filter types to obtain estimates of component velocity.

It was argued that local phase information is more robust than amplitude with respect to *1)* contrast variations, such as those caused by changes in illumination or surface orientation, and *2)* changes in scale, speed and image orientation, such as those caused by perspective projection in space-time. As a consequence, the assumptions of image translation and constant amplitude that are important for other techniques (cf. Chapter 5) can be relaxed. Chapter 6 also suggested that phase information may be viewed as a generalization of zero-crossings, but, because arbitrary values of phase are used and not just zeros, a denser set of estimates is expected, and the detection and localization of zeros is not a problem. Finally, the expression of component velocity in terms of local phase behaviour was shown to be consistent with that in terms of spatiotemporal frequency in the case of 2-d image translation.

The technique's accuracy, robustness, and localization in space-time were demonstrated in Chapter 7 through a series of experiments involving image sequences of textured surfaces moving in 3-d under perspective projection. Several of these involved sizable time-varying perspective deformation. The width of support (in all but the Yosemite sequence) was limited to 5 pixels (frames) in space (time, respectively) at one standard deviation. In most experiments with reasonably large amounts of dilation, rotation, and shear, we have found that approximately 65-80% of the component velocity estimates have errors that are less than 1°, 80-90% are within 2°, and that the proportion within 3° is generally greater than 90%. In cases dominated by image translation (Section 7.3.2) the estimates are even more accurate, often with 90% of all estimates having errors less than 1°. For a fixed aperture size, the magnitude of the errors will increase as the magnitudes of divergence, rotation, and shear increase. The experiments in Sections 7.3 and 7.4 used images of size 150×150, with an angular field of view of more than 50°. This spatial sampling is extremely crude in comparison with the human visual system, and the width of filter support is very large (i.e. greater than 1° at one standard deviation). With denser spatial sampling, and therefore smaller support functions (relative to the angular field of view), greater geometric deformation can be tolerated since the determining factors are the changes in speed, orientation or scale within the local aperture.

Chapter 8 demonstrated the accuracy with which 2-d velocity might be computed from the component velocity measurements using a least-squares fit of the component velocity estimates in each local neighbourhood to a linear model of 2-d velocity. The resultant 2-d velocity estimates are also very accurate, with most estimates within 2° of the correct 2-d velocity. As shown in Figure 8.1, a speed error of 2° amounts to relative velocity errors of 6% – 10%. This sub-pixel accuracy compares favourably with the approaches of Heeger (1988) (see Section 8.3.2), Duncan and Chou (1988) who report relative errors of 20% on realistic images, Little et al. (1988)

whose technique is limited to integer pixel velocities per frame, Waxman et al. (1988) who claim relative errors of 10%, and with the results of the differential and matching methods reported in [Little and Verri, 1989]. The results reported in Section 8.3.3 with the *Hamburg Taxi Sequence* compare favourably with those previously reported in [Enkelmann, 1986; Nagel, 1983; Nagel and Enkelmann, 1986].

To conclude this part of the monograph, we restate several important properties of the phase-based approach: First, the use of phase information is image-independent in that specific features or tokens are not a prerequisite. As a consequence, problems associated with tokens, such as detection, localization, descriptive richness, and matching, are avoided. By comparison, with zero-crossings there remain questions concerning the robustness of detection and localization [Jenkin and Jepson, 1988], and, as noted by Waxman et al. (1988) and Duncan and Chou (1988), there are also problems in areas of high edge density. Second, the present approach differs from most others in that, because of the initial representation, the image structure is separated according to velocity and scale, so that multiple velocity estimates are allowed in local neighbourhoods. This may be useful in cases of transparency, specular reflections, and partial occlusion. Third, the computational scheme discussed in Section 6.2.1 is efficient, and suitable to parallel processing. Each velocity-tuned channel may be handled independently, and the various operations are local, mainly linear, shift-invariant and separable in space-time.

With respect to the initial filters we note the following properties:

- The technique is not strictly limited to the use of separable Gabor kernels. Other band-pass filters can be used, such as inseparable kernels with non-unit aspect ratios as long as they exhibit constant phase properties.

  Furthermore, the use of phase information does not depend strictly on velocity-tuned filters. For example, motion-sensitive cortical cells do not appear to exhibit velocity-tuning in the sense discussed in Chapter 3. Their tuning appears to be separable in space-time, with a relatively large temporal frequency pass-band in comparison to a narrow, oriented, spatial frequency pass-band [Movshon et al., 1985; Grzywacz and Yuille, 1990]; they are band-pass, but exhibit only very crude velocity tuning. These could also be used as precursors to the use extraction of phase information. Other properties of the filters are discussed in Chapter 9.

- The filters should remove mean illumination and attenuate low frequencies to avoid aliasing in the subsampled representation. For Gabor filters this implies a relatively small bandwidth.

- The initial representation based on the filter outputs is efficient because it is subsampled at a reasonable rate and quantized. It is hoped that with better forms of interpolation even lower sampling rates can be tolerated with similar or better accuracy.

Finally, the least-squares method of computing 2-d velocity from component velocity is not an integral part of our approach. For it tacitly assumes that all estimates of component velocity in each local region arise from the motion of a single surface in the scene. This was discussed in the context of the Hamburg Taxi Sequence (Section 8.3.3) where it was evident that estimates of component velocity that were due to more than one object were merged into incorrect 2-d velocity estimates (in these cases the systems were often well-conditioned and produced a small residual). The occurrence of occlusion, shadows, specular reflections and transparency all cause some form of spatiotemporal intensity distortion and give rise to multiple local velocities. It is incorrect to assume that these effects will be negligible with respect to the least-squares minimization (as would white Gaussian noise), thereby producing a single, legitimate, 2-d velocity estimate. Further research is necessary to develop better approaches to these problems.

Part III

On Phase Properties of Band-Pass Signals

There currently exist several different approaches to phase-based image matching and velocity measurement. In addition to the technique developed in Chapter 6 for measuring image velocity from the phase gradient, there are several other phase-based methods for matching between two views. These include phase-difference methods [Burt et al., 1989; Jenkin and Jepson, 1988; Jepson and Jenkin, 1989; Langley et al., 1990; Sanger, 1988; Weng, 1990], phase-correlation methods [Kuglin and Hines, 1975; Girod and Kuo, 1989], and cepstrum methods [Olson and Potter, 1989; Yeshurun and Schwartz, 1989]. As discussed in Section 6.4.3 and in more detail in Chapter 11, phase-based matching defines disparity as the shift necessary to align the phase values of band-pass filtered versions of the two signals [Jenkin and Jepson, 1988; Langley et al., 1990; Sanger, 1988; Weng, 1990]. A disparity predictor can be constructed using the difference in phase between the two band-pass filtered views at a point, divided by the average instantaneous frequency of the filter responses. The predictor produces estimates of disparity, and may be used iteratively to converge to accurate matches of the local phase signals of the two views.

Although these techniques produce encouraging experimental results, we lack a satisfying explanation for their success. The usual justification for phase-based methods has rested on *a)* the Fourier shift theorem, *b)* a model of image translation between different views, and *c)* the assumption that the phase of the output of band-pass filters behaves linearly as a function of spatial position. The extent to which these techniques produce accurate measurements when there are deviations from image translation, and the extent to which phase is typically linear has not been addressed in detail. Chapter 6 claimed that phase has the important property of being stable with respect to small geometric deformations of the input that occur with perspective projections of 3-d scenes. Figures 6.1 – 6.4 clearly showed that amplitude is sensitive to geometric deformation, but no concrete justification was given for the stability of phase.

Chapter 9 discusses this issue of phase stability in the restricted case of 1-d signals, the relevant deformations of which are translations and dilations. Also discussed is the extent to which phase can be expected to be linear as a function of space and time. Using quantitative measures of phase behaviour it is shown that both phase stability and linearity depend on the form of the filters and their frequency bandwidths. These quantitative measures can also be used to predict the performance of phase-based methods as a function of the geometric deformations expected between views, and the form of the filters.

But phase can sometimes be unstable. This is clear from Figures 6.2 – 6.4 where in certain regions the phase of the filter output is very sensitive to small shifts in time or space. It is important that the form of this instability be understood and a method for its detection developed so that unreliable measurements can be discarded. For example, the implementation discussed in Chapter 7 involved constraints on the estimated local frequency and the amplitude of the filter output in order to eliminate unreliable estimates. These constraints were justified experimentally only in that they appeared to remove most of the incorrect measurements. Chapter 10 helps to justify these constraints by describing a major cause of phase instability, namely, the occurrence of phase singularities. It describes the existence of phase singularities and the regions about them in which phase is sensitive to changes in scale and spatial position. It also presents a simple method for the detection of the singularity neighbourhoods.

Most of the theoretical results concerning phase stability and instability that are developed in Chapters 9 and 10 are based on a model of white noise for the input. Chapter 11 discusses the results in the context of natural images and the measurement of binocular disparity, for which matching between the left and right views of a scene along epipolar lines is viewed as a 1-d signal matching problem. This helps to illustrate the issues of phase stability and instability in the vicinities of salient image features, the importance of instantaneous frequency, and the need to detect regions of instability due to phase singularities.

Most of the work discussed in Part III was originally reported in [Jepson and Fleet, 1990; Fleet and Jepson, 1991; Fleet et al., 1991].

Chapter 9

Scale-Space Phase Stability

This Chapter shows that phase is typically stable with respect to small geometric deformations, and is quasi-linear as a function of spatial position. These issues are addressed in the restricted case of 1-d signals, where the relevant deformations are translations and dilations like those that occur between left and right views of stereo pairs of images; scale variations between left and right binocular views of a smooth surface are often as large as 20% [Ogle, 1956]. Using a scale-space framework, we simulate changes in the scale of the input by changing the tuning of a band-pass filter. In this context our concerns include the extent to which phase is stable under small perturbations of the filter tuning, and the extent to which phase is generally linear through space. These properties are shown to depend on the form of the filters and their frequency bandwidths; for a given type of filter, larger bandwidths provide better phase stability but poorer phase linearity through space. Therefore, when one designs filters for phase-based disparity measurement, it is important that the bandwidth be set according to the expected magnitude of deformation between left and right views. Situations in which phase becomes unstable or leads to inaccurate matching are discussed at length in the next chapter.

9.1 Scale-Space Framework

In what follows we consider band-pass zero-phase filters with complex-valued kernels $K(x, \lambda)$, the real and imaginary parts of which form quadrature pairs (Hilbert transforms of one another).[1] Let $\lambda > 0$ denote a scale parameter

---

[1]Linear phase is required to implement causal temporal temporal filters. Also, note that the results of this Chapter do not hold only for quadrature-pair kernels. However, orthogonality of the real and imaginary parts, and identical amplitude spectra are natural objectives to ensure that the phase gradient is independent of absolute phase (so that phase winds smoothly about the origin in the complex plane).

that determines the frequency pass-band to which the filter is tuned. Also, let the kernels be normalized such that

$$\| K(x, \lambda) \| = 1, \tag{9.1}$$

where $\| K(x, \lambda) \|^2 \equiv\ <K(x, \lambda), K(x, \lambda)>$, which is defined by

$$<f(x), g(x)> = \int_{-\infty}^{\infty} f(x)^* g(x)\, dx, \tag{9.2}$$

where f^* denotes the complex conjugate of f. For convenience, we also assume translational invariance and self-similarity across scale (i.e., wavelets [Mallat, 1989]), so that $K(x, \lambda)$ satisfies

$$K(x, \lambda) = \frac{1}{\sqrt{\lambda}} K(x/\lambda, 1). \tag{9.3}$$

Wavelets are convenient since, because of their self-similarity, their octave bandwidth remains constant, independent of the scale to which they are tuned. The results of this chapter extend to filters other than wavelets, such as windowed Fourier transforms for which the spatial extent of the effective kernels is independent of λ.

The convolution of $K(x, \lambda)$ with an input $I(x)$ is typically written as

$$S(x, \lambda) = K(x, \lambda) * I(x). \tag{9.4}$$

Because $K(x; \lambda) \in \mathbb{C}$, the response $S(x, \lambda) = \text{Re}[S(x, \lambda)] + i\,\text{Im}[S(x, \lambda)]$ is also complex-valued, and can be expressed using amplitude and phase (1.4) as in $S(x, \lambda) = \rho(x, \lambda)\, e^{i\,\phi(x, \lambda)}$, where

$$\rho(x, \lambda) = |S(x, \lambda)|, \quad \phi(x, \lambda) = \arg[S(x, \lambda)] \in (-\pi, \pi]. \tag{9.5}$$

This chapter has two main concerns. The first is the expected stability of phase under small scale perturbations of the input. If phase is not stable under scale variations between different views, then the phase-based matching will *not* provide useful measurements of disparity or velocity. To demonstrate this, we exploit the filters' self-similarity by simulating changes in the scale of the input by changing the scale tuning of the filter: If one signal is a translation and dilation of another,

$$I_0(a(x)) = I_1(x), \tag{9.6}$$

where $a(x) = a_0 + a_1 x$, then the responses $S_0(x, \lambda_0)$ and $S_1(x, \lambda_1)$ will satisfy

$$\sqrt{a_1}\, S_0(a(x), \lambda_0) = S_1(x, \lambda_1), \quad \lambda_1 = \lambda_0 / a_1. \tag{9.7}$$

In terms of disparity measurement, if filters tuned to λ_0 and λ_1 were applied to $I_0(x)$ and $I_1(x)$, then the structure extracted by both filters would be similar (up to a scalar multiple) and the output signals would be related by precisely the same deformation $a(x)$. However, in practice we apply the *same* filters to $I_0(x)$ and $I_1(x)$ because the scale factor a_1 that relates two views is unknown. For phase-matching to yield accurate estimates of $a(x)$, the phase of the filter output should be insensitive to small scale variations of the input.

The second issue addressed in this Chapter is the extent to which phase is linear with respect to spatial position. Linearity affects the ease with which the phase signal can be differentiated in order to estimate the instantaneous frequency of the filter response. Linearity also affects the speed and accuracy of disparity measurement based on phase-difference disparity predictors; if the phase signal is exactly linear, then the disparity can be computed in just one step, without requiring iterative refinement (see Chapter 11).

For illustrative purposes, let $K(x, \lambda)$ be a Gabor kernel[2] [Gabor, 1946], $Gabor(x, \sigma(\lambda), k(\lambda))$, where

$$Gabor(x; \sigma, k) = G(x; \sigma) e^{ixk} \tag{9.8}$$

$$G(x; \sigma) = \frac{1}{(\sqrt{\pi}\sigma)^{1/2}} e^{-x^2/2\sigma^2} . \tag{9.9}$$

The peak tuning frequency of the Gabor filter is given by

$$k(\lambda) = \frac{2\pi}{\lambda} , \tag{9.10a}$$

and, with the extent of the Gaussian envelope measured at one standard deviation, and a bandwidth of β octaves, the radius of spatial support (4.10) is

$$\sigma(\lambda) = \frac{1}{k(\lambda)} \left(\frac{2^\beta + 1}{2^\beta - 1} \right) . \tag{9.10b}$$

Figure 9.1 *(top)* shows a signal composed of a sample of white Gaussian noise concatenated with a scanline from a natural image. The two middle images show the amplitude and phase components of the scale-space Gabor response, $\rho(x, \lambda)$ and $\phi(x, \lambda)$, over two octaves. The bandwidth of the Gabor filter was 0.8 octaves. In these scale-space plots, spatial position is shown on the horizontal axis and log scale is shown on the vertical axis. The bottom images show level contours of $\rho(x, \lambda)$ and $\phi(x, \lambda)$. In the context of the scale-space expansion, an image property is said to be *stable* where its level contours are vertical. Figure 9.1 shows that $\rho(x, \lambda)$ depends significantly on scale as its level contours are not generally vertical. By contrast, the

---

[2]Strictly speaking the real and imaginary Gabor parts do not have identical amplitude spectra. They are however orthogonal, and for sufficiently small bandwidths (e.g. an octave or less), they are very close to being in quadrature.

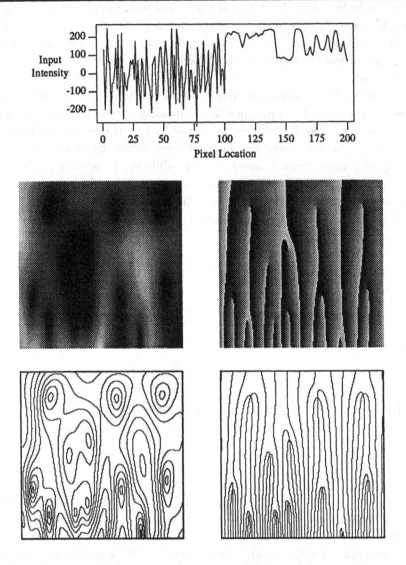

Figure 9.1. **Gabor Scale-Space Expansion:** *The input signal (top) consists of a sample of white Gaussian noise (left side), and a scanline from a natural image (right side). The middle images show the amplitude and phase components of $S(x, \lambda)$ generated by a Gabor filter with a bandwidth of $\beta = 0.8$ octaves, with $12 \leq \lambda \leq 48$ pixels (on a log scale). The horizontal and vertical axes represent spatial position and log scale. The bottom two figures show level contours of constant $\rho(x, \lambda)$ and $\phi(x, \lambda)$.*

phase structure is generally stable, except for several isolated regions to be discussed below. Gradient-based techniques applied to low-pass or band-pass filtered images produce inaccurate velocity estimates, in part, because they track level contours of the filter output, implicitly requiring that *both* the amplitude and the phase be stable.

The response $S(x, \lambda)$ defined in (9.4) is referred to as a scale-space expansion of $I(x)$. It is similar to band-pass expansions defined by the Laplacian of a Gaussian ($\nabla^2 G$), but it is expressed in terms of complex-valued filters (cf. [Koenderink, 1984; Mallat, 1989; Witkin, 1981; Yuille and Poggio, 1984]). Zero-crossings of $\nabla^2 G$ output are equivalent to crossings of constant phase of a complex band-pass filter, the real part of which is $\nabla^2 G$. Here we use scale-space to investigate the effects of small perturbations of the input scale on image properties that might be used for matching. Note that we are not proposing a new multi-scale image representation, nor are we interpreting phase behaviour in terms of specific image features such as edges.

9.2 Kernel Decomposition

The stability/linearity of phase can be examined in terms of the behaviour of the phase difference between an arbitrary scale-space location and other points in its neighbourhood. To see this, let $S_j \equiv S(x_j, \lambda_j)$ denote the filter response at scale-space position $\mathbf{p}_j = (x_j, \lambda_j)$, and for convenience, let S_j be expressed using inner products instead of convolution:

$$S_j \;=\; <K_j^*(x), I(x)> \;, \tag{9.11}$$

where $K_j(x) \equiv K(x - x_j, \lambda_j)$. Phase differences in the neighbourhood of an arbitrary point $\mathbf{p}_0$ as a function of relative scale-space position $\mathbf{p}_1 - \mathbf{p}_0 = (\Delta x, \Delta \lambda)$ can be written as

$$\Delta \phi(\mathbf{p}_0, \mathbf{p}_1) \;=\; \arg[S_1] - \arg[S_0] \;. \tag{9.12}$$

Phase is perfectly stable when $\Delta\phi$ is constant as a function of $\Delta\lambda$, and it is linear with respect to spatial position when $\Delta\phi$ is a linear function of Δx.

To examine the behaviour of $S(x, \lambda)$ and $\Delta\phi$ in the neighbourhood of $\mathbf{p}_0$ we write the scale-space response at $\mathbf{p}_1$ in terms of S_0 and a residual term R_1 that goes to zero as $\| \mathbf{p}_1 - \mathbf{p}_0 \| \to 0$; that is,

$$S_1 \;=\; z_1 S_0 + R_1 \;. \tag{9.13}$$

Equation (9.13) is easily derived if the effective kernel at $\mathbf{p}_1$, that is $K_1(x)$, is written as the sum of two orthogonal terms, one in the span of $K_0(x)$ and the other orthogonal to $K_0(x)$; that is,

$$K_1(x) \;=\; z_1 K_0(x) + H_1(x) \;, \tag{9.14}$$

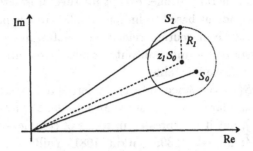

Figure 9.2. **Sources of Phase Variation:** *This shows the formation of S_1 in terms of S_0, the scalar multiple z_1, and the additive residual R_1.*

where z_1 and $H_1(x)$ are given by

$$z_1 = <K_0(x), K_1(x)> , \tag{9.15}$$

$$H_1(x) = K_1(x) - z_1 K_0(x) . \tag{9.16}$$

Equation (9.13) follows from (9.14) with $R_1 = <H_1^*(x), I(x)>$. The complex scalar z_1 reflects the cross-correlation of the two kernels K_0 and K_1. The behaviour introduced by R_1 is related to the signal structure to which K_1 responds but K_0 does not. Both z_1 and H_1 are functions of relative scale-space position $(\Delta x, \Delta \lambda)$.

Equation (9.13), depicted in Figure 9.2, shows that the phase of S_1 can differ from the phase of S_0 in two ways: *1)* the additive phase shift due to z_1; and *2)* the phase shift caused by the additive residual term R_1. Phase will be *stable* under small scale perturbations whenever both the phase variation due to z_1 as a function of scale and $|R_1|/|z_1 S_0|$ are reasonably small. If $|R_1|/|z_1 S_0|$ is large then phase remains stable only when R_1 is in phase with S_0, that is, if $\arg[R_1] \approx \arg[S_0]$. Otherwise, phase may vary wildly as a function of either spatial position or small scale perturbations.

9.3 Phase Stability Given White Gaussian Noise

To characterize the stability of phase behaviour through scale-space we first examine the response of $K(x, \lambda)$ to stationary, white Gaussian noise. Using the kernel decomposition (9.14), we derive approximations to $E[\Delta \phi]$ and $E[|\Delta \phi - E[\Delta \phi]|]$, where $E[\cdot]$ denotes mathematical expectation. The mean phase difference provides a prediction for the phase behaviour, and the expected variation of the phase difference about the mean amounts to our confidence in the prediction. These approximations can be shown to depend only on the cross-correlation z_1 (9.15), and are outlined below; they are derived in further detail in Appendix E.

Given white Gaussian noise, the two signals R_1 and $z_1 S_0$ are independent, because $<z_1 K_0(x), H_1(x)> = 0$, and the phase of S_0 is uniformly distributed over $(-\pi, \pi]$. If we also assume that $\arg[R_1]$ and $\arg[z_1 S_0]$ are uncorrelated and that $\arg[R_1] - \arg[z_1 S_0]$ is uniform over $(-\pi, \pi]$, then the residual signal R_1 does not affect the mean phase difference. That is, we can approximate $E[\Delta\phi]$ by

$$\tilde{\mu}(z_1) = \arg[z_1], \qquad (9.17)$$

where z_1 is a function of scale-space position. The component of $\Delta\phi$ about the approximate mean is then given by (cf. Figure 9.2)

$$\begin{aligned}\tilde{\Delta\phi} &= \Delta\phi - \tilde{\mu}(z_1) \\ &= \arg[z_1 S_0 + R_1] - \arg[z_1 S_0] . \qquad (9.18)\end{aligned}$$

The expected magnitude of $\tilde{\Delta\phi}$ provides a measure of the spread of the distribution of $\Delta\phi$ about the mean.[3] With the assumptions that $|R_1| < |z_1 S_0|$ and that $\arg[R_1]$ is uniformly distributed, it is shown in Appendix E that an approximate bound, $\tilde{\sigma}(z_1)$, on $E[\,|\tilde{\Delta\phi}|\,]$ is given by the ratio of $|R_1|$ and $|z_1 S_0|$. This bound is also shown to reduce to

$$\tilde{\sigma}(z_1) = \frac{\sqrt{1 - |z_1|^2}}{|z_1|} . \qquad (9.19)$$

9.3.1   Gabor Kernels

For $K(x, \lambda) = Gabor(x; \sigma(\lambda), k(\lambda))$, z_1 is given by (see Appendix F.2)

$$z_1 = \sqrt{2\pi}\, G(\Delta x; \bar{\sigma})\, G(\Delta k; \frac{\bar{\sigma}}{\sigma_0 \sigma_1})\, e^{i\,\Delta x\,[k_0 + (\Delta k\, k_0^2 / \bar{k}^2)]} \qquad (9.20)$$

where $k_j = k(\lambda_j)$ defines the filter tunings (9.10a), $\sigma_j = \sigma(\lambda_j)$ defines the support widths (9.10b), $\Delta k = k_1 - k_0$, $\bar{k} = \sqrt{k_0^2 + k_1^2}$, and $\bar{\sigma} = \sqrt{\sigma_0^2 + \sigma_1^2}$.

The approximate mean phase difference, that is $\arg[z_1]$, is given by

$$\tilde{\mu}(z_1) = \Delta x \left(k_0 + \frac{\Delta k\, k_0^2}{\bar{k}^2} \right) . \qquad (9.21)$$

---

[3]The expected value of the $|\tilde{\Delta\phi}|$ is one measure of the spread of the probability density function. Compared to the standard deviation it is more robust with respect to outliers [Huber, 1981], and in this case, it yields analytic results while the second moment does not (see Appendix E).

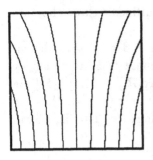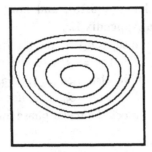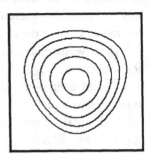

Figure 9.3. Phase Stability with Gabor Kernels: *Scale-space phase behaviour near* $p_0 = (0, \lambda_0)$ *is shown with log scale on the vertical axis over two octaves with* $\lambda_1 \in (\frac{1}{2}\lambda_0, 2\lambda_0)$, *and spatial position on the horizontal axis,* $x_1 \in (-\lambda_0, \lambda_0)$. *(left) Level contours of* $\tilde{\mu}(z_1) = n\pi/2$, *for* $n = -4, ..., 4$, *as a function of scale-space position. These contours are independent of the bandwidth. (middle and right) Level contours of* $\tilde{\sigma}(z_1)$ *are shown for* $\beta = 0.8$ *and* 1.0. *In each case, contours are shown for* $\tilde{\sigma}(z_1) = 0.3n$ *where* $n = 1, ..., 5$; *the innermost contours correspond to* $\tilde{\sigma}(z_1) = 0.3$.

Figure 9.3 (left) shows level contours of $\tilde{\mu}(z_1)$ for a Gabor filter. The point $p_0 = (0, \lambda_0)$ lies at the centre. Log scale and spatial position in the neighbourhood of p_0 are shown on vertical and horizontal axes. These contours illustrate the expected mean phase behaviour, which has the desired properties of stability through scale and linearity through space. The contours are essentially vertical near p_0, and for fixed Δk, the mean phase behaviour $\tilde{\mu}(z_1)$ is a linear function of Δx. It is also interesting to note that the mean phase behaviour does not depend significantly on the bandwidth of the filter; for Gabor filters it is evident from (9.21) that $\tilde{\mu}(z_1)$ is independent of β.

The expected magnitude of $\Delta \phi$ about the mean (9.19) can also be determined from (9.20) straightforwardly. In particular, from (9.19), we expect $\tilde{\sigma}(z_1)$ to behave linearly in the neighbourhood of p_0, because for sufficiently small $\| (\Delta x, \Delta \lambda) \|$ it can be shown that $|z_1| = 1 + O(\| (\Delta x, \Delta \lambda) \|^2)$. The middle and right panels of Figure 9.3 show level contours of $\tilde{\sigma}(z_1)$ for Gabor kernels with bandwidths β of 0.8 and 1.0 octaves. In both cases, $\tilde{\sigma}(z_1)$ is monotonically decreasing as one approaches p_0 in the centre. As $\| p_1 - p_0 \|$ decreases so does the relative magnitude of R_1, and therefore so does expected fluctuation in the phase of S_1 about the phase of $z_1 S_0$. By design, $\tilde{\sigma}(z_1)$ is a measure of the distribution of phase differences about the mean. The contours in Figure 9.3 show that the expected deviation from $\tilde{\mu}(z_1)$ is small near p_0. Since $\tilde{\mu}(z_1)$ has the properties we desire (stability through

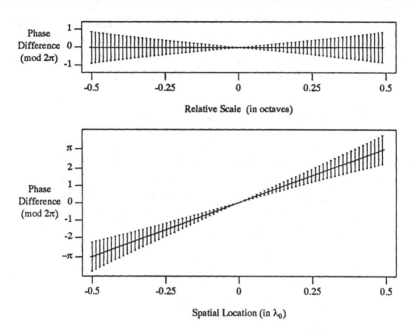

Figure 9.4. **Behaviour of $\Delta\phi$ for 1-D Slices of Scale-Space:** *The expected behaviour of $\Delta\phi$ is shown for vertical and horizontal slices through the middle of the scale-space in Figure 9.3 for Gabor filters with $\beta = 1$. Both plots show the approximate mean $\tilde{\mu}(z_1)$ and the variation about the mean $\tilde{\sigma}(z_1)$ (confidence bars). (top) Phase behaviour as a function of $\Delta\lambda$ while $\Delta x = 0$. (bottom) Phase behaviour as a function of Δx for $\Delta\lambda = 0$.*

scale and linearity through space near $\mathbf{p_0}$), we can also view $\tilde{\sigma}(z_1)$ as a direct measure of phase stability and phase linearity.

An alternative demonstration of this behaviour is shown in Figure 9.4. The top panel shows $\tilde{\mu}(z_1)$ with confidence bars ($\tilde{\sigma}(z_1)$) as a function of scale for a vertical slice through the Gabor scale-space with $\beta = 1$. The bottom panel shows $\tilde{\mu}(z_1)$ and $\tilde{\sigma}(z_1)$ as a function of spatial position for a horizontal slice through the same scale-space expansion. Observe the constant and linear behaviour of the mean in the two cases.

It is also evident from Figure 9.3 *(middle and right)* that $\tilde{\sigma}(z_1)$ depends on the bandwidth of the filter. This follows from the dependence of σ_0 and σ_1 in (9.20) on bandwidth as given by (9.10b). As the bandwidth increases phase becomes stable for greater scale perturbations of the input, while the spatial extent over which phase is generally linear appears to decrease. The smallest contours in Figure 9.3 *(middle and right)* encircle $\mathbf{p_0} = (0, \lambda_0)$ and correspond to $\tilde{\sigma}(z_1) = 0.3$ which amounts to a phase difference of about $\pm$ 5% of a wavelength. For $\beta = 1$ this contour encloses approximately $\pm$ 20% of an octave vertically and $\pm$ 20% of a wavelength spatially. In this case, relative scale changes of 10% and 20% are typically accompanied by phase shifts of less than 3.5% and 6.6% of a wavelength respectively.

9.3.2   Verification of Approximations

There are other ways to illustrate the behaviour of phase differences as a function of scale-space position. Although they do not yield much explanatory insight they help to validate the approximations discussed above.

First, following Davenport and Root (1958) it can be shown that the probability density function for $\Delta\phi$, at a given scale-space position, for a quadrature-pair kernel and white Gaussian noise input is

$$pdf(\Delta\phi) \;=\; \frac{2\pi\,A\left[(1 - B^2)^{1/2} + B(\pi - \arccos B)\right]}{(1 - B^2)^{3/2}}\,, \qquad (9.22)$$

for $-\pi \leq \Delta\phi < \pi$, where A and B are given by

$$A \;=\; \frac{s_0^2 s_1^2 - c_1^2 - c_2^2}{4\pi^2\,s_0^2 s_1^2}\,, \qquad B \;=\; \frac{c_1 \cos\Delta\phi + c_2 \sin\Delta\phi}{s_0 s_1}\,,$$

$s_0 = \;\| \operatorname{Re}[K_0(x)] \|$, $\; s_1 = \;\| \operatorname{Re}[K_1(x)] \|$, $\; c_1 = \;<\operatorname{Re}[K_0(x)],\, \operatorname{Re}[K_1(x)]>$, and $\; c_2 = \;<\operatorname{Re}[K_0(x)],\, \operatorname{Im}[K_1(x)]>$.

Given the density function, we can numerically integrate to find its mean behaviour and the expected variation about the mean. Figure 9.5 shows the behaviour of the mean *(left)* and the absolute variation about the mean *(middle)* of $\Delta\phi$ as a function of scale-space position for Gabor kernels with a bandwidth of 0.8 octaves. Figure 9.5 *(right)* shows the expected variation of $\Delta\phi$ about the mean for Gabor filters of 1.0 octave. The mean behaviour in this case is not shown as it is almost identical to Figure 9.5 *(left)*. In all three cases level contours have been superimposed to better illustrate the behaviour. Intensity in Figure 9.5 *(left)* reflects values between $-\pi$ and π. Values in the other two range from 0 in the centre to $\pi/2$ at the edges where the distribution of $\Delta\phi$ becomes close to uniform. Comparing Figures 9.3 and 9.5, notice that the bound in (9.19) is tightest for smaller values of $\Delta\phi$ (see Appendix E); the two smallest contours in both figures are extremely close.

Finally, we compare the phase behaviour predicted by $\tilde{\mu}(z_1)$ and $\tilde{\sigma}(z_1)$ with actual statistics of phase differences gathered from scale-space expansions of different input signals. Figure 9.6 *(top)* shows behaviour of $\Delta\phi$ predicted by $\tilde{\mu}(z_1)$ and $\tilde{\sigma}(z_1)$ for one octave Gabor filters as in Figure 9.4. The lower two panels in Figure 9.6 show statistical estimates of $\mathrm{E}[\Delta\phi]$ and $\mathrm{E}[\|\Delta\phi - \mathrm{E}[\Delta\phi]\|]$ measured from scale-space expansions of white noise and of

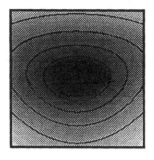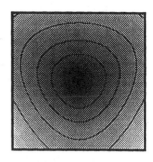

Figure 9.5. $\mathbf{E}[\Delta\phi]$ and $\mathbf{E}[\,|\Delta\phi-\mathbf{E}[\Delta\phi]|\,]$ **for White Noise Inputs:** *Scale-space phase behaviour based on (9.22) for Gabor kernels. As above, $\mathbf{p}_0 = (0, \lambda_0)$ is the centre, with $-\lambda_0 \leq x_1 \leq \lambda_0$ and $\frac{1}{2}\lambda_0 \leq \lambda_1 \leq 2\lambda_0$. The same level contours as in Figure 9.3 are superimposed for comparison. (left) $E[\Delta\phi]$ for $\beta = 0.8$. (middle and right) $E[\,|\Delta\phi - E[\Delta\phi]|\,]$ for $\beta = 0.8$ and $\beta = 1.0$.*

scanlines of natural images. In both cases the observed phase behaviour is in very close agreement with the predicted behaviour. The statistical estimates of $E[|\Delta\phi - E[\Delta\phi]|]$ in these and other cases were typically only 1–5% below the bound $\tilde{\sigma}(z_1)$ over the scales shown.

9.3.3 Dependence on Filter

These quantitative measures of expected phase behaviour can be used to predict the performance of phase matching as a function of the deformation between the input signals. The same measures can also be used to examine the expected performance of different filters; for although Gabor filters (9.8) are popular [Sanger, 1988; Jepson and Jenkin, 1989; Fleet and Jepson, 1990; Langley et al., 1990; Fleet et al., 1991], several alternatives have been suggested in the context of phase information. Weng (1990) used a self-similar family of kernels derived from a square-wave window:

$$K(x, \lambda) = \begin{cases} \frac{1}{\sqrt{\lambda}} e^{ik(\lambda)x} & \text{if } |x| \leq \lambda/2 \\ 0 & \text{otherwise .} \end{cases} \tag{9.23}$$

Other alternatives, common to phase-correlation techniques, are families of filters with fixed spatial extents but tuned to different scales, also called windowed Fourier transforms [Kuglin and Hines, 1975; Girod and Kuo, 1989]. This section addresses several issues related to the choice of filter.

The effect of bandwidth on phase stability was illustrated in Figure 9.3,

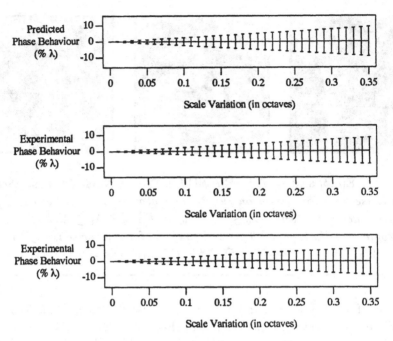

Figure 9.6. **Predicted Versus Actual Phase Behaviour:** *(top) Predicted behaviour of $\Delta\phi$ based on $\tilde{\mu}(z_1)$ and $\tilde{\sigma}(z_1)$ for a Gabor filter with $\beta = 1$. The bottom two plots show statistical estimates of $E[\Delta\phi]$ and $E[|\Delta\phi - E[\Delta\phi]|]$ that were extracted from Gabor scale-space expansions of white noise (middle) and scanlines from natural images (bottom).*

and can be explained using the form of z_1 (9.15) and Parseval's theorem:

$$z_1 \;=\; <K_0(x), K_1(x)> \;=\; (2\pi)^{-1} <\hat{K}_0(k), \hat{K}_1(k)> , \quad (9.24)$$

where $\hat{K}(k)$ denotes the Fourier transform of $K(x)$. As the bandwidth increases, the extent of the amplitude spectrum increases, and so does the range of scales $\Delta\lambda$ over which $\hat{K}_0(k)$ and $\hat{K}_1(k)$ may remain highly correlated. Similar arguments hold in space with respect to the extent of spatial support and phase linearity. For wavelets, the stability of phase with respect to input scale changes is constant across different scales, but the spatial extent over which phase is expected to be nearly linear decreases for higher scales as a function of the support width of the filters. For windowed Fourier transforms, the support width is constant for different scales, and hence so is the expected extent of linearity, but the stability of phase with respect to scale perturbations decreases at higher scales as the octave bandwidth decreases.

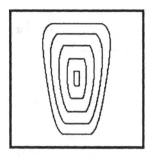  

Figure 9.7. **Modulated Square-Wave Windows:** *Scale-space phase behaviour is shown for modulated square-wave windows (9.23) in comparison to Gabor filters. (left) Level contours $\tilde\sigma(z_1) = 0.3n$ for $n = 1..., 5$ for the square-wave kernel. (middle) Similar level contours of $\tilde\sigma(z_1)$ for a comparable Gabor kernel with $\beta = 1.5$. (right) The superposition of level contours from square-wave and Gabor kernels.*

But does a simultaneous increase in the extent of the kernels' support in space and its amplitude spectrum produce better scale stability or more extensive linearity through space? For example, compared to Gaussians that *minimize* a measure of simultaneous extent (see Section 3.3), modulated square-wave kernels (9.23) have relatively large simultaneous extents in space and frequency domain. Their amplitude spectra are particularly broad, and therefore we might expect better stability than with Gabor filters.

Figure 9.7 shows the scale-space behaviour of $\tilde\sigma(z_1)$ for the modulated square-wave kernel (9.23), for which z_1 is given by (see Appendix F.2)

$$z_1 = i\,e^{i\Delta x k_1}\left(e^{i\Delta k a} - e^{i\Delta k b}\right)(\Delta k\sqrt{\lambda_0\lambda_1})^{-1}, \qquad (9.25)$$

where $\Delta\lambda$ and Δk are defined above,[4] and

$$b = \frac{\lambda_0}{2} + \min\left[0, \frac{\Delta\lambda}{2} - \Delta x\right], \qquad a = \frac{-\lambda_0}{2} + \max\left[0, -\frac{\Delta\lambda}{2} - \Delta x\right].$$

The mean phase behaviour, given by $\arg[z_1]$, is very much like that exhibited by Gabor filters in Figure 9.3 *(left)* and is not shown. But the distribution of $\Delta\phi$ about the mean, shown in Figure 9.7 *(left)*, is two to three times larger near $\mathbf{p_0}$. This suggests poorer stability *and* poorer linearity. For comparison, Figure 9.7 *(middle)* shows level contours of $\tilde\sigma(z_1)$ for a Gabor kernel with a bandwidth comparable to the modulated square-wave filter, and the superposition of the two sets of level contours *(right)*. The innermost contour of

---

[4]As $\Delta k \to 0$, this expression for z_1 converges to $\exp[i\Delta x k_0]\,(1 - \Delta x/\lambda_0)$.

the Gabor filter clearly encloses the innermost contour of the square-wave filter. This Gabor filter handles scale perturbations of 10% with an expected phase drift of up to $\pm$ 3.6% of a wavelength, while a perturbation of 10% for the modulated square-wave kernels gives $\tilde{\sigma}(z_1) = 0.47$ which amounts to a phase difference of about $\pm 7.5\%$.

The poorer phase stability exhibited by the modulated square-wave kernel implies a wider distribution of measurement errors; the measurements of velocity and disparity will not reflect the projected motion field and the projected disparity field as reliably. The poorer phase linearity affects the accuracy and speed of the disparity predictor, requiring more iterations to match the phase values between views. The poorer linearity exhibited in Figure 9.7 also contradicts the claim in [Weng, 1990] that only modulated square-wave filters produce quasi-linear phase behaviour. Wider amplitude spectra do not necessarily ensure greater phase stability. Phase stability is due to cross-correlation between kernels at different scales, that is z_1. The shapes of both the amplitude and phase spectra will therefore have significant effects. The square-wave amplitude spectra is wide, but with considerable ringing so that z_1 falls off quickly with small scale changes.

Another issue concerning the choice of filter is the ease with which phase behaviour can be accurately extracted from a subsampled encoding of the filter output. As discussed in Chapter 4, it is natural that the outputs of different band-pass filters be quantized and subsampled to avoid an explosion in the number of bits needed to represent filtered versions of the input. However, because of the aliasing inherent in subsampled encodings, care must be taken in subsequent numerical interpolation/differentiation. For example, we found that, because of the broad amplitude spectrum of the modulated square-wave and its sensitivity to low frequencies, sampling rates had to be at least twice as high as those with Gabors (with comparable bandwidths, $\beta = 1.5$) to obtain reasonable numerical differentiation.

9.4 Multiple Dimensions

Finally, although beyond the scope of the current work, it is important to note that this basic framework can be extended to consider the stability of multidimensional filters with respect to other types of geometric deformation. In particular, we are interested in the phase behaviour of 2-d oriented filters with respect to small amounts of shear and rotation as well as scale changes. This analysis can be done, as above, using the cross-correlation between a generic kernel and a series of deformations of it. In this way, quantitative approximations can be found to predict the expected degree of phase drift under different geometric deformations of the input.

Chapter 10

Scale-Space Phase Singularities

Chapter 9 discussed the stability of phase information using quantitative bounds on the expected variation of phase as a function of small scale perturbations of the input. But from Figure 9.1 it is also clear that phase stability is not uniform throughout scale-space; some regions exhibit much greater instability in that the phase contours are nearly horizontal and not vertical as desired. Because some geometric deformation is expected between left and right binocular views, or between frames of an image sequence, this instability, if not detected, is bound to cause poor measurements.

This chapter shows that an important cause of phase instability is the occurrence of phase singularities: points where the filter response passes through the origin in the complex plane. We present a qualitative model for typical phase singularities and the neighbourhoods about them in which phase is unstable. The description of singularity neighbourhoods is important because it yields a better understanding of the form of instability. It also leads to a simple method for detecting regions of instability so that unreliable measurements of binocular disparity and image velocity can be detected and discarded. This detection method helps to explain the success of the constraints used to identify unreliable velocity measurements in Chapter 7. It can also be used to improve the performance of zero-crossing and phase-based methods for computing disparity [Jenkin and Jepson, 1988; Langley et al., 1990a; Sanger, 1988; Kuglin and Hines, 1975; Girod and Kuo, 1989; Marr and Poggio, 1979; Hildreth, 1984; Waxman et al., 1988].

10.1   Occurrence of Phase Singularities

For general images $I(x)$, the scale-space defined by (9.4) is analytic and contains a number of isolated zeros, where $S(x, \lambda) = 0$.[1] Zeros of $\rho(x, \lambda)$

---

[1] Generically, where $S(x_0, \lambda_0) = 0$, the first-order terms will be non-zero with $S_x(x_0, \lambda_0) \neq S_\lambda(x_0, \lambda_0) \neq 0$.

appear as black spots in Figure 9.1. The phase signal (9.5) is also analytic, except at the zeros of $S(x, \lambda)$ where the response passes through the origin in the complex plane. This causes phase discontinuities, where the phase angle jumps by π.

In order to describe the occurrence and the effects of phase singularities, we begin with a simple model that generates singularities and exhibits the observed behaviour. It consists of two complex exponentials, the amplitudes of which vary as a function of scale. Let the input $I(x)$ be the superposition of two waveforms at frequencies k_1 and k_2, where $k_2 > k_1$. The output of a band-pass filter with peak tuning frequency $k(\lambda)$ somewhere between k_1 and k_2 can be expressed as

$$S(x, \lambda) = \rho_1(\lambda) e^{ik_1x+\psi} + \rho_2(\lambda) e^{ik_2x} , \qquad (10.1)$$

where ψ is an arbitrary phase shift. Because of the roll-off of the amplitude spectrum away from the peak tuning frequency, an increase in the scale to which the filter is tuned causes variation in the two amplitudes ρ_1 and ρ_2:

$$\frac{d\rho_1}{d\lambda} > 0 , \qquad \frac{d\rho_2}{d\lambda} < 0 . \qquad (10.2)$$

Singularities occur where $\rho_1(\lambda) = \rho_2(\lambda)$ and $k_1x + \psi = -k_2x$. That is, when the two amplitudes are identical, the singularities occur where the two signals are π radians out of phase, so that the two terms in (10.1) cancel.

Figure 10.1 shows a simple example of (10.1) in which $k_2 = 2k_1$. The four panels show the signal in the complex plane as a function of x, for successively larger ratios of ρ_1/ρ_2. The arrows indicate the direction of increasing values of x. The first panel illustrates the signal behaviour for a scale just below a singularity, that is, for $\rho_2 > \rho_1$ and $\lambda < \lambda_0$ where λ_0 denotes the scale at which the singularity occurs. The upper-right panel illustrates the case $\lambda = \lambda_0$, for which the singularity occurs where $S(x, \lambda)$ passes through the origin. The bottom two plots show the behaviour at two scales above λ_0, where ρ_1 / ρ_2 is greater than one and increasing. This sequence illustrates the general way in which higher frequency information is lost as λ increases and one moves up in scale-space.

Accordingly, we expect that the density of singularities in a region of scale-space is related to the change in the expected frequency of response from one scale to another. More precisely, the number of phase singularities in a rectangular scale-space region of width Δx, between scales λ_0 and $\lambda_0 + \Delta\lambda$, should be proportional to the width Δx and to the change in the number of times the signals wind about the origin in the complex plane over the distance Δx. If the expected winding number over Δx is taken to be $\Delta x / \lambda$ where λ is the scale of the filter tuning, then we expect that the approximate number of singularities in a rectangular region of size $\Delta x \Delta\lambda$, and near scale λ_0, is proportional to $\Delta x \Delta\lambda / \lambda_0^2$; they occur reasonably often.

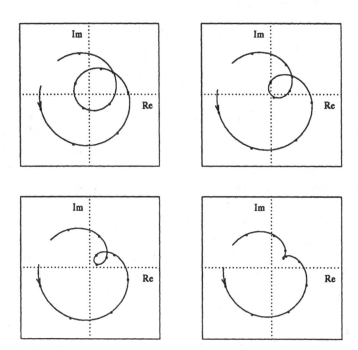

Figure 10.1. **Simple Model of Singularity:** *The evolution of phase through a singularity is shown using (10.1) with four different ratios of ρ_1/ρ_2 (corresponding to four increasing values of λ). The bullets mark fixed spatial positions, and the arrows show the direction of increasing values of x.*

10.2 Qualitative Model of Singularity Neighbourhoods

This simple model (10.1) helps to illustrate the cause of phase singularities as well as the signal behaviour near singular locations in scale-space. Our goal throughout the rest of Chapter 10 is to explain in some detail the form of phase instability that occurs in the scale-space neighbourhoods of phase singularities, and to characterize these regions so that we may detect them.

To begin we consider properties of the complex logarithm of the filter response $S(x, \lambda)$; that is, $\log S(x, \lambda) \equiv \log \rho(x, \lambda) + i\phi(x, \lambda)$, where $\rho(x, \lambda)$ and $\phi(x, \lambda)$ are the amplitude and phase components of $S(x, \lambda)$. In particular, observe that the x-derivative of $\log S$ satisfies

$$\frac{\partial}{\partial x} \log S(x, \lambda) = \frac{[S^*(x, \lambda)\, S_x(x, \lambda)\,]}{|S(x, \lambda)|^2}$$

$$= \frac{\rho_x(x, \lambda)}{\rho(x, \lambda)} + i\,\phi_x(x, \lambda)\,. \qquad (10.3)$$

The imaginary part of (10.3), $\tilde{k} \equiv \phi_x(x, \lambda)$, is the instantaneous frequency of S, and the real part, $\tilde{\rho} \equiv \rho_x(x, \lambda)\,/\,\rho(x, \lambda)$, is the relative amplitude derivative. Our characterization of the singularity neighbourhoods will be given in terms of the behaviour of $\tilde{k}$ and $\tilde{\rho}$. Detailed support for this description is given in Sections 10.4 and 10.5.[2] In what follows, let (x_0, λ_0) denote the location of a typical singularity.

The neighbourhoods just above and below singular points can be characterized in terms of the behaviour of $\phi_x(x, \lambda)$. Above singular points, where $\lambda > \lambda_0$, they are characterized by $\tilde{k} \ll k(\lambda)$; that is, the local frequencies are significantly below the corresponding peak tuning frequencies. Within these neighbourhoods there exist *retrograde regions* where local frequencies are negative, that is $\tilde{k} < 0$. In Figure 10.1 *(bottom-left)*, where the signal normally winds counter-clockwise about the origin, the retrograde behaviour corresponds to the middle of the small loop where the signal winds clockwise with respect to the origin. Along the boundaries of retrograde regions (which begin and terminate at singular points) the local frequency is zero; that is, $\phi_x(x, \lambda) = 0$. The significance of this is that, where $\phi_x(x, \lambda) = 0$, the level phase contours are horizontal in scale-space and not vertical as desired. Nearby this boundary, both inside and outside the retrograde region, the level contours are generally far from vertical, which, as discussed above, implies considerable phase instability.

Below singular points, where $\lambda < \lambda_0$, the neighbourhoods are characterized by $\tilde{k} \gg k(\lambda)$; that is, the local frequencies are significantly greater than the corresponding peak tuning frequencies. In addition, the local frequency changes rapidly as a function of spatial location. This implies numerical instability in a subsampled signal representation in these neighbourhoods.

We illustrate this behaviour using the scale-space in Figure 9.1; Figure 10.2 shows three 1-d slices of $\phi(x, \lambda)$ and $\phi_x(x, \lambda)$ at scales $\lambda = 41$, 38, and 20. The signals at $\lambda = 41$ and 38 pass just above and just below three phase singularities respectively. The negative frequencies occurring just above the singular points, and the high positive frequencies occurring just below the singular points are evident in these two plots. The bottom slice

---

[2]The computation of $\tilde{k}$ and $\tilde{\rho}$ follows straightforwardly from (G.1) and (G.2) in Appendix G and from the discussion in Appendix C concerning the numerical interpolation and differentiation of the subsampled filter output.

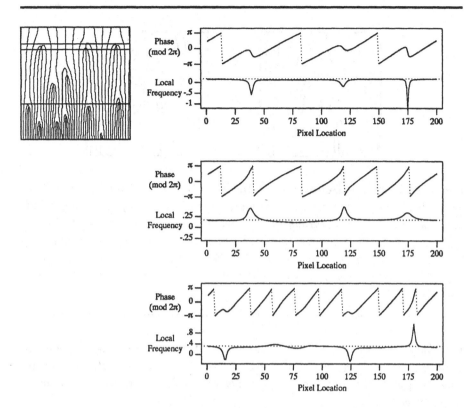

Figure 10.2. **Phase and Local Frequency Near Singularities:** *(left) Level contours from $\phi(x, \lambda)$ of Figure 9.1 are shown with three scales marked with solid horizontal lines (at $\lambda = 41$, 38, and 20). (right) Plots of $\phi(x, \lambda)$ and $\phi_x(x, \lambda)$ for each scale. The first lies just above 3 singular points while the second lies just below the same singular points. The third passes near 3 different singularities. Vertical dotted lines denote phase wrapping (not discontinuities), and the horizontal dotted line marks the filters' peak tuning frequencies $k(\lambda) = 0.154, 0.165,$ and 0.314 respectively.*

passes through three singularity neighbourhoods, just above two singularities (near locations 17 and 124), and just below one singularity (near location 180). Between locations 55 and 85 the effects of two distant singularities are evident as the local frequency rises and then dips slightly.

Figure 10.3 *(left)* shows the typical behaviour of level phase contours near a singularity, taken from the scale-space expansion of a scanline from a real image. The phase singularity is the point in the middle through which several of the phase contours pass. The small elliptical contour shows the retrograde

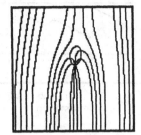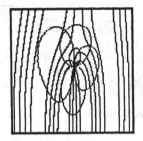

Figure 10.3. **Phase Behaviour in Singularity Neighbourhoods:** *(left) Typical behaviour of level phase contours near a singularity. The singularity is the point in the centre through which several phase contours pass. The small ellipsoidal contour identifies points at which $\phi_x(x, \lambda) = 0$. The singularity lies at the bottom of this contour. (right) The added contours show the regions delineated by the constraints used to detect the singularity neighbourhoods (cf. Section 10.3).*

boundary where $\phi_x(x, \lambda) = 0$. The instability above the singular point is clear from the nearly horizontal level phase contours. Directly below the singular point, the high local frequencies are evident from the high density of phase contours.

Finally, the scale-space neighbourhoods to the left and the right of phase singularities can be characterized in terms of amplitude variation. As we approach a singular point, $\rho(x, \lambda_0)$ goes to zero. Based on a simple linear model of amplitude, at a specific location x_1 near the singularity, the distance to the singular point is approximately $|\Delta x| = \rho(x_1, \lambda_0)/|\rho_x(x_1, \lambda_0)|$. Equivalently, as we approach the singularity $|\rho_x(x_1, \lambda_0)|/\rho(x_1, \lambda_0)$ increases. As a consequence, the neighbourhoods to the left and right of phase singularities can be characterized by large values of the relative amplitude derivative $|\rho_x(x_1, \lambda_0)|/\rho(x_1, \lambda_0)$.

10.3 Detecting Singularity Neighbourhoods

To use phase information for the measurement of image velocity or binocular disparity, we must first detect singularity neighbourhoods so that measurements within them, which are unreliable, may be discarded. Here we introduce constraints on local frequency and amplitude that enable us to identify locations within singularity neighbourhoods, while avoiding the explicit localization of the singular points. It is important that this can be done using information at only one scale, so that a dense set of samples across many scales is unnecessary.

The neighbourhoods above and below the singular points can be detected using the distance between the instantaneous frequency $\phi_x(x, \lambda)$ and the peak tuning frequency $k(\lambda)$, as a function of the radius of the amplitude spectrum (measured at one standard deviation $\sigma_k(\lambda)$):[3]

$$\frac{|\phi_x(x, \lambda) - k(\lambda)|}{\sigma_k(\lambda)} \leq \tau_k, \tag{10.4}$$

where $\sigma(\lambda) = 1/\sigma_k(\lambda)$ is given by (9.10b). For example, Figure 10.4 *(top-left)* shows the regions of the scale-space in Figure 9.1 that were detected by this constraint with 3 different values of the threshold τ_k. As explained in more detail in Section 10.5, the level contours of (10.4) for specific values of τ_k form 8-shaped regions that increase in size as the values of τ_k decrease.

The neighbourhoods adjacent to singular points can be detected with a local amplitude constraint:

$$\sigma(\lambda) \frac{|\rho_x(x, \lambda)|}{\rho(x, \lambda)} \leq \tau_\rho, \tag{10.5}$$

where $\sigma(\lambda)$ defines the radius of filter support. The regions detected by this constraint are ∞-shaped and increase in size as τ_ρ decreases (see Figure 10.4 *(top-right)*).

Finally, we can combine (10.4) and (10.5) into a single constraint to detect entire singularity neighbourhoods:

$$\left| \frac{\partial}{\partial x} \log S(x, \lambda) - ik(\lambda) \right| \leq \sigma_k(\lambda)\tau. \tag{10.6}$$

This constraint detects a roughly ellipsoidal region about each singular point that increases in size as τ decreases. Figure 10.4 *(middle-left)* shows the regions detected by (10.6) for three values of τ. The final three panels in Figure 10.4 show the original phase contours from Figure 9.1, the contours that remain after the application of the stability constraint (10.6) with $\tau = 1.25$, and the contours within the regions detected by (10.6). This constraint typically detects about 15–20% the scale-space area.

10.4   Singularity Theory

The goal of the next two sections is to provide further support for the characterization of, and the detection method for, the singularity neighbourhoods

---

[3]This constraint is similar to the frequency constraint (7.3) that was used in Chapter 7 to detect unreliable velocity measurements.

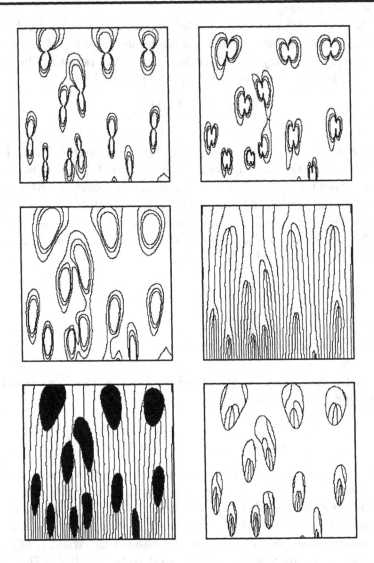

Figure 10.4. Detection of Singularity Neighbourhoods: *This shows the application of the stability constraints to the scale-space in Figure 9.1. (top) Level contours of (10.4) and (10.5) with τ_k and τ_ρ set to 1.0, 1.25 and 1.5. (middle-left) Level contours of (10.6) with $\tau = 1.0, 1.25$, and 1.5. (middle-right) Level phase contours corresponding to Figure 9.1. (bottom) The phase contours not detected by (10.6) and those in regions detected by (10.6) with $\tau = 1.25$.*

that were outlined in Sections 10.2 and 10.3. This section shows that, given the appropriate kernels, the magnitude of the instantaneous frequency increases like $|\lambda - \lambda_0|^{-1}$ as one approaches the location of a singularity from above or below in scale-space, and the magnitude of the relative amplitude derivative increases like $|x - x_0|^{-1}$ as the singularity is approached from the side. Furthermore, it is shown that extreme negative and positive frequencies typically occur above and below the singular point respectively.

Let $\mathbf{p}_0$ be the location of a singularity, so that $S(x_0, \lambda_0) = 0$. Since $S(x, \lambda)$ is analytic, the response S_1 in the neighbourbood of the singularity can be expressed as a Taylor series approximation about $\mathbf{p}_0$:

$$S(x_1, \lambda_1) = \Delta x\, S_x(x_0, \lambda_0) + \Delta \lambda\, S_\lambda(x_0, \lambda_0) + O\left((\Delta x + \Delta \lambda)^2\right) \quad (10.7)$$

where

$$S_x(x_0, \lambda_0) = <K_x^*(x - x_0, \lambda_0), I(x)> , \quad (10.8a)$$

$$S_\lambda(x_0, \lambda_0) = <K_\lambda^*(x - x_0, \lambda_0), I(x)> . \quad (10.8b)$$

Thus, for $\mathbf{p}_1$ sufficiently close to $\mathbf{p}_0$, the effective kernel is approximately $\Delta x\, K_x(x - x_0, \lambda_0) + \Delta \lambda\, K_\lambda(x - x_0, \lambda_0)$. For certain kernels, such as Gabor functions with support widths that remain constant as a function of scale, and for general images (so that $S_x \neq S_\lambda \neq 0$), it is possible to use (10.7) and (10.8) to specify the behaviour of instantaneous frequency and the relative amplitude derivative in the neighbourhood of a singular point.

For example, in the case of Gabor kernels with constant linear bandwidth where the support width σ does not depend on scale λ, as in $K(x, \lambda) = e^{ixk(\lambda)}G(x; \sigma)$, the derivatives of $K(x, \lambda)$ evaluated at $\mathbf{p}_0$ satisfy

$$K_\lambda(x - x_0, \lambda_0) = i\frac{2\pi\sigma^2}{\lambda_0^2}K_x(x - x_0, \lambda_0) . \quad (10.9)$$

From (10.7), with (10.8) and (10.9), the response at $\mathbf{p}_1$ can be approximated, to first-order, by

$$S(x_1, \lambda_1) \approx \left(\Delta x + i\frac{2\pi\sigma^2}{\lambda_0^2}\Delta \lambda\right) S_x(x_0, \lambda_0) . \quad (10.10)$$

Since S_x in (10.7) is constant (independent of $\mathbf{p}_1$), the first-order structure of $S(x, \lambda)$ in the neighbourhood of $\mathbf{p}_0$ is determined by $(\Delta x + i\,(2\pi\sigma^2 / \lambda^2)\,\Delta\lambda)$. From (10.3), the x-derivative of the complex logarithm of the response is given by

$$\frac{\partial}{\partial x}logS(x_1, \lambda_1) \approx \frac{\Delta x - i\,\kappa\,\Delta\lambda}{\Delta x^2 + \kappa^2\,\Delta\lambda^2} , \quad (10.11)$$

where $\kappa = 2\pi\sigma^2 / \lambda_0^2$. Therefore, as $\|(\Delta x, \Delta\lambda)\|$ decreases to zero, the magnitude of $\frac{\partial}{\partial x}logS(x_1, \lambda_1)$ increases like $\|(\Delta x, \kappa\,\Delta\lambda)\|^{-1}$.

From (10.11) and (10.3), the local frequency and relative amplitude derivative in the case of Gabor kernels with constant linear bandwidth near a singularity are given by

$$\tilde{k} \approx \frac{-\kappa\,\Delta\lambda}{\Delta x^2 + \kappa^2\,\Delta\lambda^2}\,, \tag{10.12a}$$

$$\tilde{\rho} \approx \frac{\Delta x}{\Delta x^2 + \kappa^2\,\Delta\lambda^2}\,. \tag{10.12b}$$

Because $\mathrm{sgn}[\tilde{k}] = -\mathrm{sgn}[\Delta\lambda]$, the local frequency will be positive for $\lambda_1 < \lambda_0$, and negative for $\lambda_1 > \lambda_0$. This is consistent with the qualitative model given in Section 10.2. Also note that, as we approach the singular point from directly above or below (i.e. with $\Delta x = 0$), the magnitude of local frequency increases like $|\Delta\lambda^{-1}|$ (towards ∞) as $\Delta\lambda \to 0$. Similarly, $\mathrm{sgn}[\tilde{\rho}] = \mathrm{sgn}[\Delta x]$, and for $\Delta\lambda = 0$ the magnitude of the relative amplitude derivative increases like $|\Delta x^{-1}|$ as $\Delta x \to 0$. Thus, this analysis supports the characterization of the immediate neighbourhood about a singularity for a specific class of band-pass filters.

Unfortunately, this form of analysis is somewhat limited for two reasons. First, it is very sensitive to the exact form of the kernel. Other kernels, such as Gabor functions with constant octave bandwidths (or wavelets), may exhibit almost the same behaviour but lack the exact analytic properties. With the discrete approximations to kernels that are used in practice, the singularity theory is overly precise. Second, it is only appropriate in the immediate neighbourhood of a singularity. To describe the scale-space behaviour further from phase singularities the analysis becomes cumbersome, requiring that higher-order structure be taken into account. Yet such a broad description is necessary to explain the nature of phase instabilities in singularity neighbourhoods like those shown in Figures 10.2 – 10.4. We take a different approach in the next section to overcome some of these difficulties.

10.5   Phase Singularities and Gaussian Noise

Based on a Gaussian noise model for the input, the responses of nearby points in scale-space were shown in Chapter 9 to be highly correlated. Using a similar approach we can examine the expected behaviour in the neighbourhood of a singularity, thereby supporting the characterization and the detection of singularity neighbourhoods described in Sections 10.2 and 10.3 for a wider class of kernels (like those defined in Chapter 9). We do this by considering the mean behaviour of $\tilde{k}$ and $\tilde{\rho}$ when the input consists of white Gaussian noise.

In order to model $S(x, \lambda)$ in the neighbourhood of a singularity we use the fact that the real and imaginary parts of S_0 are both zero at the singular point $\mathbf{p}_0$; that is,

$$< \mathrm{Re}[K_0(x)], \, I(x) > \quad = \quad < \mathrm{Im}[K_0(x)], \, I(x) > \quad = \quad 0 \, . \qquad (10.13)$$

Analogous to the development in Section 9.2, the response S_1 at a nearby point $\mathbf{p}_1$ can be viewed as the output of a new kernel $\tilde{H}_1(x)$ consisting of $K_1(x)$ projected onto the orthogonal complement of the real and imaginary parts of $K_0(x)$. We can generate $\tilde{H}_1(x)$ explicitly using Gram-Schmidt orthogonalization; we project out the components of $K_1(x)$ in the span of $\mathrm{Re}[K_0(x)]$ and $\mathrm{Im}[K_0(x)]$ to create $\tilde{H}_1$;[4]

$$\mathrm{Re}[\tilde{H}_1(x)] \;=\; \mathrm{Re}[K_1(x)] \;-\; \frac{< \mathrm{Re}[K_1(x)], \, \mathrm{Re}[K_0(x)] >}{\| \, \mathrm{Re}[K_0(x)] \, \|^2} \mathrm{Re}[K_0(x)]$$
$$-\; \frac{< \mathrm{Re}[K_1(x)], \, \mathrm{Im}[K_0(x)] >}{\| \, \mathrm{Im}[K_0(x)] \, \|^2} \mathrm{Im}[K_0(x)]$$

$$(10.14a)$$

$$\mathrm{Im}[\tilde{H}_1(x)] \;=\; \mathrm{Im}[K_1(x)] \;-\; \frac{< \mathrm{Im}[K_1(x)], \, \mathrm{Re}[K_0(x)] >}{\| \, \mathrm{Re}[K_0(x)] \, \|^2} \mathrm{Re}[K_0(x)]$$
$$-\; \frac{< \mathrm{Im}[K_1(x)], \, \mathrm{Im}[K_0(x)] >}{\| \, \mathrm{Im}[K_0(x)] \, \|^2} \mathrm{Im}[K_0(x)] \, .$$

$$(10.14b)$$

Given $\tilde{H}_1(x)$ we may investigate the behaviour of $S(x, \lambda)$, its instantaneous frequency, and the relative amplitude derivative in the neighbourhood of a singularity. This helps to demonstrate the properties of retrograde regions and the effects of different filter properties such as bandwidth.

Let the input be mean-zero, white Gaussian noise. With this assumption we can determine probability density functions for the local frequency $\tilde{k}(x_1, \lambda_1)$ and the relative amplitude derivative $\tilde{\rho}(x_1, \lambda_1)$, based on the kernels $\tilde{H}_1(x)$ and $\tilde{H}_1' \equiv \frac{d}{dx} \tilde{H}_1(x)$.[5] In Appendix G these are shown to be

$$pdf(\tilde{k}) \;=\; |\tilde{k}| \, A \int_0^{-\pi} \frac{\sin^2 \psi \; d\psi}{(s_1^2 \, \tilde{k}^2 \;+\; (s_2^2 - 2c_2 \tilde{k}) \sin^2 \psi \;-\; c_1 \tilde{k} \sin 2\psi \,)^2} \, ,$$

$$(10.15a)$$

$$pdf(\tilde{\rho}) \;=\; |\tilde{\rho}| \, A \int_{\pi/2}^{-\pi/2} \frac{\cos^2 \psi \; d\psi}{(s_1^2 \, \tilde{\rho}^2 \;+\; (s_2^2 - 2c_1 \tilde{\rho}) \cos^2 \psi \;-\; c_2 \tilde{\rho} \sin 2\psi \,)^2} \, ,$$

$$(10.15b)$$

where $A = (s_1^2 s_2^2 - c_1^2 - c_2^2)/\pi$, $s_1 = \| \, \mathrm{Re}[\tilde{H}_1(x)] \, \|$, $s_2 = \| \, \mathrm{Re}[\tilde{H}_1'(x)] \, \|$, $c_1 = < \mathrm{Re}[\tilde{H}_1(x)], \, \mathrm{Re}[\tilde{H}_1'(x)] >$, and $c_2 = < \mathrm{Re}[\tilde{H}_1(x)], \, \mathrm{Im}[\tilde{H}_1'(x)] >$.

---

[4]Note that the kernels $\tilde{H}_1(x)$ are not identical to the kernels $H_1(x)$ given by (9.16). It can be shown, for instance, that $< H_1(x), \mathrm{Re}[K_0] > \neq 0$, so that some of the real and imaginary components of K_0 do exist in $H_1(x)$, yet should be zero near a singularity. We must therefore explicitly project out both the real and imaginary parts individually.

[5]Here, $\tilde{H}_1'(x)$ is constructed from $K_1'(x)$ using the Gram-Schmidt procedure as in (10.14), by projecting out the real and imaginary parts of $K_0(x)$ from $K_1'(x)$.

Given a specific kernel, either in analytic form or the discrete set of coefficients, we can numerically approximate the expected behaviour of $\tilde{k}$ and $\tilde{\rho}$ based on (10.12) through (10.15). For example, for Gabor kernels with bandwidths of one octave, Figure 10.5 shows mean frequency and mean amplitude derivative as functions of scale-space position about a singular point. In both cases the singular point $\mathbf{p}_0 = (x_0, \lambda_0)$ is at the centre, wavelengths are shown for 40% of an octave above and below λ_0, and $-0.4\,\lambda_0 \leq \Delta x \leq 0.4\,\lambda_0$. The level contours are added to the intensity to help illustrate the behaviour of $\tilde{k}$ and $\tilde{\rho}$ in these neighbourhoods. The expected response frequency at scale λ_0 in Figure 10.5 *(left)* away from a singularity is $\tilde{k} \approx 0.6$. The level contours corresponding to values of $\tilde{k}$ somewhat larger than 0.6 form smaller circular loops just below $\mathbf{p}_0$, that begin and terminate at $\mathbf{p}_0$. The larger the value of $\tilde{k}$, the smaller the loop. The level contours corresponding to smaller values of $\tilde{k}$ form circular loops above the singular point. In particular, the second smallest such contour corresponds to $\tilde{k} = 0$, thereby marking the expected retrograde boundary. In Figure 10.5 *(right)* the level contours of $\tilde{\rho}$ also form circular loops, but they lie to the sides of the singular point. The smaller loops correspond to larger values of $|\tilde{\rho}|$.

It is also informative to take a closer look at the expected behaviour of $\tilde{k}$ directly above and below $\mathbf{p}_0$ as a function of $\Delta\lambda$ with $\Delta x = 0$. If $\Delta x = 0$, then it is easy to show that $c_1 = 0$ in (10.15). The probability density function for $\tilde{k}$ (10.15a) can then be shown to simplify to

$$p(\tilde{k}) \;=\; \frac{s_1 s_2 - c_2^2}{2 s_1 \left(s_1^2\, \tilde{k}^2 + s_2^2 - 2 c_2 \tilde{k} \right)^{3/2}}\,. \qquad (10.16)$$

Furthermore, the probability of negative frequencies, which is the probability of falling within a retrograde region, can be shown to be

$$\mathrm{prob}(\tilde{k} < 0) \;=\; \frac{1}{2}\left(1 - \frac{c_2}{s_1 s_2}\right)\,. \qquad (10.17)$$

Observe that $c_2 / s_1 s_2$ is the cross-correlation coefficient for $\mathrm{Re}[\tilde{H}_1(x)]$ and $\mathrm{Im}[\tilde{H}_1'(x)]$. High positive correlation implies little chance of negative frequencies, while negative correlation implies low frequencies with significant chance of negative frequencies. This interpretation is also clear from the expression of local frequency in (G.1) in Appendix G.

Figure 10.6 shows numerical approximations to the cross-correlation coefficient $c_2 / s_1 s_2$ and $\mathrm{prob}(\tilde{k} < 0)$, for a vertical slice through scale-space (as a function of $\Delta\lambda$, with $\Delta x = 0$), for each of three different kernels. In each case, the singular point corresponds to a wavelength of $\lambda_0 = 1$. The top plot illustrates the case of Gabor kernels with constant support width (i.e. constant linear bandwidth). In this case the standard deviation was set so that the bandwidth at λ_0 was one octave. As suggested in Section 10.4 by

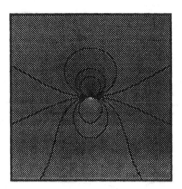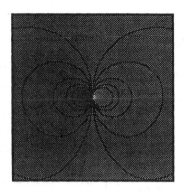

Figure 10.5. **Mean Frequency and Relative Amplitude Derivative in Singularity Neighbourhoods:** *The mean instantaneous frequency $E[\tilde{k}]$ and the mean relative amplitude derivative $E[\tilde{\rho}]$ are shown as a function of scale-space position. Their probability density functions are given by (10.15), and the singular point $\mathbf{p}_0 = (x_0, \lambda_0)$ is at the centre of each figure. Also, in both cases, $-0.4\,\lambda_0 \leq \Delta x \leq 0.4\,\lambda_0$, and wavelengths are shown for 40% of an octave above and below λ_0. The level contours are shown for $E[\tilde{k}] = 0.25\,n$ where $n = -1, 0, ..., 6$, and for $|E[\tilde{\rho}]| = 0.1, 0.25, 0.5, 0.75, 1.0$. (The intensities were compressed by $\sqrt{E[\tilde{k}]}$ and $\sqrt{E[\tilde{\rho}]}$ for visibility, since both quantities behave like d^{-2} where d is the distance from the singularity.)*

(10.12a), the probability of negative local frequency jumps from 0 just below $\mathbf{p}_0$ to 1 just above $\mathbf{p}_0$. The middle plot in Figure 10.6 used Gabor functions with bandwidths fixed at one octave. It shows that there is a non-zero probability of retrograde behaviour just below a singularity, because $\mathrm{prob}(\tilde{k} < 0)$ is not zero for $\lambda < \lambda_0$, and it is not 1 for $\lambda > \lambda_0$. It is clear, however, that with high probability the behaviour is very similar to that exhibited in Figure 10.6 *(top)*. The bottom plot in Figure 10.6 used Gabor functions with bandwidths of 0.8 octaves. The main difference between this behaviour and the behaviour in the case of 1 octave bandwidths is the expected extent of the retrograde region above the singularity, which can be defined as the smallest distance $\lambda_1 - \lambda_0$ such that $\mathrm{prob}(\tilde{k}(\lambda_1) < 0) = 0.5$.

Chapter 9 showed that the range of scales over which phase is generally stable and the extent of the spatial region in which phase is typically close to linear are determined in part by the bandwidth of the filters. As the bandwidth increases, the scale stability is enhanced while the spatial extent of linear phase behaviour decreases. Similarly, for singularity neighbourhoods, as the bandwidth increases, the vertical extent of the singular-

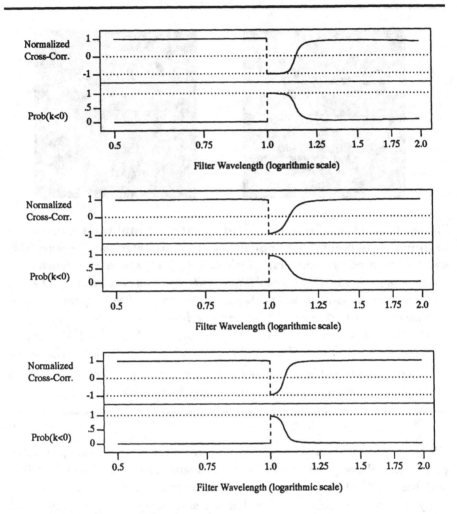

Figure 10.6. **Retrograde Behaviour Above and Below Singularities:** *Three examples of $c_2 / s_1 s_2$ and $prob(\tilde{k} < 0)$ (10.17) are shown for a vertical slice through scale-space (as a function of $\Delta \lambda$, with $\Delta x = 0$). (top) For Gabor functions with the constant support width σ. (middle) For Gabor functions with $\beta = 1$. (bottom) For Gabor functions with $\beta = 0.8$.*

ity neighbourhoods also increases while the spatial extent of the singularity neighbourhoods decreases. An interesting question for future study concerns the relation between filter bandwidth and the area in scale-space occupied by singularity neighbourhoods. Ignoring other constraints, it is desirable to choose a bandwidth that minimizes the size of singularity neighbourhoods.

Chapter 11

Application to Natural Images

The results of Chapters 9 and 10 concerning the stability of phase information were derived mainly with a white Gaussian noise model of the input. We now consider in more detail their application to real images. The issue of phase stability is discussed in Section 11.1 in the context of natural images that contain salient image features, such as edges, and regions with virtually no structure at all to which the filters respond. We then illustrate the results in the context of binocular disparity, for which matching band-pass filtered versions of the left and right views of a scene along epipolar lines can be viewed as a 1-d problem. After discussing the basic approach in Section 11.2, we consider the accuracy of phase-based disparity measurements for 1-d signals with and without the removal of singularity neighbourhoods, and then as a function of scale and translation perturbations between two views. Section 11.4 briefly illustrates some of these points with 2-d images.

11.1 Phase Stability/Instability

Unlike white noise, the Fourier harmonics of natural images are often correlated across scales, and their amplitude spectra typically decay like $1/k$ [Field, 1987]. Both of these facts affect our results concerning phase stability, the accuracy of phase matching, and instability due to singularity neighbourhoods.

First, because of the decay of the input amplitude spectra the filter responses will generally be biased to lower frequencies (as compared to white noise). As a result, care is required to ensure that the filter outputs do not contain too much power at low frequencies. Otherwise, there will likely be a) more distortion due to aliasing in a subsampled representation of the response, and b) larger singularity neighbourhoods, and hence a sparser set of reliable measurements. These problems are evident when comparing the modulated square-wave filters with Gabor filters because the former has greater sensitivity to low frequencies in many cases.

Second, without the assumption of white noise we should expect $z_1 S_0$ and R_1 in (9.13) to be correlated. This may lead to improved phase stability when R_1 and S_0 remain in phase, and poorer stability when they become systematically out of phase. Although we lack a sufficient model of natural images, in terms of local structure, to provide a detailed treatment of phase stability on general images, several observations are readily available, and suggest that phase is typically stable, especially in the neighbourhoods of salient image features. The filter output in a given region is essentially a weighted set of harmonics. In the vicinity of image features such as edges, bars and ramps, we expect greater phase stability because the phases of the input harmonics (unlike the white noise) are already coincident. This is clear from their Fourier transforms. Therefore changing scales slightly, or adding new harmonics at the high or low end will not change the phase of the output very much.

It is also worth noting that this phase coincidence at the feature locations coincides with local maxima of the amplitude response [Morrone and Burr, 1988]. When different harmonics are in phase their amplitudes combine additively. When out of phase they will cancel one another. Therefore it can be argued that neighbourhoods of local amplitude maxima correspond to regions in which phase is maximally stable (as long as the signal-to-noise-ratio is sufficiently high). This is independent of the absolute phase at which the different harmonics coincide,[1] and is significant for stable phase-based matching.

As one moves away from salient features, such as edges, the different harmonics may become increasingly out of phase, responses from different features may interfere, and the amplitude of response decreases. This yields two main types of instability: *1)* where interference between the responses to several nearby features causes the total response to disappear at isolated points; and *2)* where large regions have very small amplitude and are dominated by noise. The first case amounts to a phase singularity and is detectable using the stability constraint (10.6). In the second case the phase behaviour in different views may be dominated by uncorrelated noise, but *not* violate the stability constraint. For these situations a signal-to-noise constraint is necessary.

To illustrate these points, Figure 11.1 shows the Gabor scale-space expansion of a signal containing several step edges (Figure 11.1 *(top)*). Moving left-to-right and top-to-bottom, the first two images show the amplitude response

---

[1] Morrone and Burr (1988) argue that psychophysical salience correlates well with phase coincidence only for certain absolute values of phase, namely, integer multiples of $\pi/2$, which are perceived as edges and bars of different polarities.

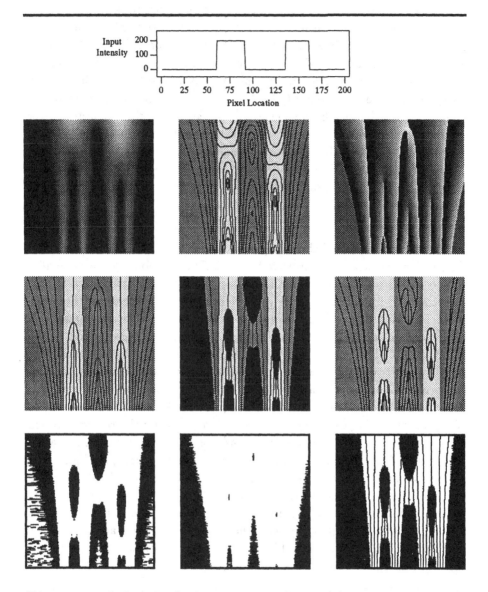

Figure 11.1. **Gabor Scale-Space Expansion With Step-Edge Input:**
The input signal (top) consists of two bars (30 and 25 pixels wide, 45 pixels apart). $S(x, \lambda)$ was generated by Gabor filters with $\beta = 1$. The vertical and horizontal axes represent log scale (over 2 octaves $12 \leq \lambda \leq 48$) and spatial position. The level contour plots have been superimposed upon the input to show the relative location of the edges. Descriptions of the individual images are given in the text.

and its level contours superimposed on the input signal (replicated through scale) to show the relationship between amplitude variation and the edges. The next four images show the scale-space phase response, its level contours, the contours that remain after the detection of singularity neighbourhoods using (10.6) with $\tau = 1.25$, and the phase contours in the neighbourhoods detected by (10.6). As expected, phase is stable near the edge locations where the local Fourier harmonics are in phase, and $\arg[R_1] \approx \arg[S_0]$ over a wide range of scales. The similarity of the interference patterns between the different edges to the singularity neighbourhoods discussed in Chapter 10 is clear, and these regions are detected using the stability constraint. Also detected are the regions relatively far from the edges where the amplitude and phase responses of the filter both go to zero.

However, as explained above, regions in which the filter response decreases close to zero are also very sensitive to noise. These regions can become difficult to match since uncorrelated noise between two views can dominate the response. To illustrate this, we generated a different version of the scale-space in which uncorrelated noise was added to the input independently before computing each scale (to simulate uncorrelated noise added to different views). The response to the independent noise patterns satisfied the stability constraint much of the time, but the phase structure was unstable (uncorrelated) between scales. Figure 11.1 *(bottom-left)* shows the regions detected by the stability constraint in this case. As discussed, the regions of low amplitude in the original are now dominated by the response to the noise and can no longer be detected. Another constraint on the signal-to-noise ratio of the filter output appears necessary. For example, Figure 11.1 *(bottom-middle)* shows the regions in which the amplitude of the filter output is 5% or less of the maximum amplitude at that scale, similar to the constraint used in Section 7.2 to detect unreliable velocity measurements. This constraint in conjunction with those discussed earlier are sufficient to obtain phase stability comparable to that in the noiseless case (Figure 11.1 *(bottom-right)*).

This also demonstrates the fact that an amplitude constraint alone does not serve our purposes, for the constraints on instantaneous frequency and amplitude derivative detect different regions. Although the regions detected by a single threshold applied directly to $\rho(x, \lambda)$ will eventually include the regions detected by (10.6) if the threshold is large enough, they will also enclose regions of stable phase behaviour. A single amplitude constraint will remove more of the signal than necessary if relied on to detect all of the instabilities.

11.2 Measurement of Binocular Disparity

We now apply the results of the last two chapters to the problem of measuring binocular disparity. We begin with a description of a phase-difference method for binocular matching.

If the geometric configuration of two cameras is known, then the *scene-point disparity* (i.e. the relative distance between the projections of a single scene point onto the two imaging surfaces) encodes the 3-d depth of the point. By convention, assume that crossed disparities, which correspond to scene features nearer than the fixation point, are negative. Uncrossed disparities, which correspond to scene features farther away than the fixation point, are positive. The computation of *image-point disparity* (or simply, image disparity) involves the matching of image features in one view with those in the other in order to derive an approximation to scene-point disparity.

Let $I_l(x)$ and $I_r(x)$ denote the left and right input signals, and let $R_l(x)$ and $R_r(x)$ denote the left and right responses of a complex-valued band-pass filter with peak tuning frequency k_0:

$$R_l(x) = \rho_l(x)\,e^{i\,\phi_l(x)}\,, \qquad R_r(x) = \rho_r(x)\,e^{i\,\phi_r(x)}\,. \qquad (11.1)$$

Following [Jenkin and Jepson, 1988] and [Sanger, 1988], the problem of computing binocular disparity can be formulated in terms of phase matching, that is, determining the shift required to equalize the phases of the left and right signals. The local image disparity, at a specific position x, for an initial guess d_i, is defined to be the shift $d(x)$ such that

$$\phi_l\left(x - \frac{d(x)}{2}\right) = \phi_r\left(x + \frac{d(x)}{2}\right)\,, \qquad (11.2)$$

and $|d(x) - d_i|$ is as small as possible. Because of phase periodicity, this approach can be expected to deal only with shifts of less than half a wavelength to either side of the initial guess. If the disparity is too large (i.e. if the local wavelength is too small with respect to the shift) then the computed phase difference can be wrong by a multiple of 2π. This yields incorrect disparity measurements. The use of filters tuned to higher frequencies, with relatively large disparities, therefore requires a control strategy such as coarse-to-fine propagation [Nishihara, 1984; Jenkin and Jepson, 1988; Sanger, 1988; Langley et al., 1990]. Here we concentrate solely on the basic disparity measurements, while assuming that the initial guess is sufficiently close to the true disparity; the control strategy is beyond the scope of this monograph.

When the initial guess is sufficiently good and the filter outputs are sinusoidal with constant frequency k_0, the shift necessary to match the phases of the left and right filter outputs is given by (cf. [Jenkin and Jepson, 1988;

Sanger, 1988])

$$\tilde{d}_0(x) \equiv \frac{[\phi_l(x) - \phi_r(x)]_{2\pi}}{k_0} , \qquad (11.3)$$

where $[\theta]_{2\pi} \in (-\pi, \pi]$ denotes the principal part of θ. However, the disparity estimates provided by $\tilde{d}_0$ will not be exact if the outputs are *not* sinusoidal with constant frequency k_0. For example, if the left and right filter outputs are shifted versions of one another with disparity s, then the phase difference in the numerator of (11.3) becomes

$$\Delta\phi(x) = \phi\left(x + \frac{s}{2}\right) - \phi\left(x - \frac{s}{2}\right) . \qquad (11.4)$$

If $\phi(x)$ is smooth (singularities were discussed in Chapter 10), we can rewrite (11.4) with $\phi(x)$ expressed as a Taylor series about x as $\Delta\phi(x) = s\,\phi'(x) + O(s^3\,\phi'''(x))$. The disparity error, $\epsilon(x) = d(x) - \tilde{d}_0(x)$, with $\tilde{d}_0(x) = \Delta\phi(x)/k_0$ as in (11.3), can now be written as

$$\begin{aligned}
\epsilon(x) &= s - \tilde{d}_0(x) \\
&= s\left(1 - \frac{\phi'(x)}{k_0}\right) + O\left(\frac{s^3\,\phi'''(x)}{k_0}\right) . \qquad (11.5)
\end{aligned}$$

The order-s term in (11.5) arises from the discrepancy between the peak tuning frequency k_0 and the local frequency of the filter output $\phi'(x)$.

This technique can be improved by adopting a more general model in which k_0 in (11.3) is replaced by the average local frequency between the left and right views. This yields

$$\tilde{d}_1(x) = \frac{[\phi_l(x) - \phi_r(x)]_{2\pi}}{\bar{k}(x)} , \qquad (11.6)$$

where $\bar{k}(x) = (\phi_l'(x) + \phi_r'(x))/2$. Then, $\tilde{d}_1(x)$ yields the exact disparity when the left and right views both have frequencies that are locally constant, but not necessarily equal to the filter tuning k_0. When the left and right outputs are shifted versions of one another, but do not have constant frequency, the disparity error $\epsilon(x) = s - \tilde{d}_1(x)$ for (11.6) reduces to

$$\epsilon(x) = O\left(\frac{s^2\,\phi'''(x)}{\phi'(x)}\right) . \qquad (11.7)$$

Notice that (11.7), in contrast to (11.5), contains no order-s term.

In any case, if the disparity estimates are not exact, shifting the signals by $\tilde{d}_1(x)$ will not precisely match the phases in the left and right filter outputs as required by (11.2). However, because the error depends on the initial disparity (i.e. s in (11.5) and (11.7)), one way to improve the accuracy is to iterate the basic measurements. On one cycle of the iteration the images are first shifted according to the current disparity approximation, the local image disparity between the shifted left and right signals is computed using either $\tilde{d}_0$ or $\tilde{d}_1$, and the disparity approximation is updated with the result. That is, with an initial guess d_i, the disparity approximation at iteration $t + 1$ is given by

$$\tilde{d}^{t+1}(x) = \tilde{d}^t(x) + \Delta\tilde{d}^t(x), \tag{11.8}$$

where

$$\tilde{d}^0(x) = d_i,$$

$$\Delta\tilde{d}^t(x) = \frac{[\phi_l(x - \tilde{d}^t(x)/2) - \phi_r(x + \tilde{d}^t(x)/2)]_{2\pi}}{k(x)},$$

and $k(x)$ is given by $\bar{k}(x)$ in (11.6) for the new predictor, or by k_0 for the old predictor (11.3). The estimate $\tilde{d}^t(x)$ converges to the exact disparity, as defined by (11.2). With bandwidths of one octave, the local frequency may be up to $k_0/2$ away from k_0, in which case the first-order errors in (11.5) can be as large as $s/2$. Therefore, an upper bound on the convergence rate for the old disparity predictor $\tilde{d}_0(x)$ is 1 bit per iteration. The new predictor $\tilde{d}_1(x)$ converges quadratically.

11.3 Experiments: Dilation and Translation

To demonstrate the improved disparity estimates that result from the predictor $\tilde{d}_1(x)$ in (11.6), and the importance of singularity neighbourhoods, we have done experiments with affine deformations between left and right views. The data for most of the experiments were collected from scanlines from real images (results from white noise input were similar in most respects). In Section 11.4 the technique is applied to a pair of real images in contrast to the technique of Jepson and Jenkin (1989).

In the experiments described below, we used Gabor filters with bandwidths of 0.8 octaves and principal wavelengths between 4 and 100 pixels. For illustrative purposes some outputs were over-sampled (as in Figure 9.1) but all computations used an effective subsampling rate of one (complex) sample every $\sigma(\lambda)$, where $\sigma(\lambda)$ denotes the standard deviation of the Gaussian window. The computation of $\frac{d}{dx}R(x)$ that is required to compute the local frequency and the relative amplitude derivative was based on the same computation as described in Sections 6.2.1 and 7.1.

Figure 11.2 clearly shows the errors that occur in singularity neighbourhoods. In this case the left and right phase signals were shifted versions of a slice of the scale-space (see Figure 10.2 *(bottom)* with $\lambda = 20$) shown in

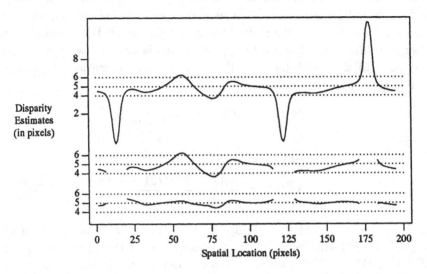

Figure 11.2. **Disparity Measurement:** *The top two plots show the disparity estimates of $\tilde{d}_0(x)$ before and after the removal of the singularity neighbourhoods. The bottom plot shows the results of the improved predictor $\tilde{d}_1(x)$,
with the singularity neighbourhoods removed. The tuned wavelength λ was 20
pixels and the disparity (error in the initial guess) was 5 pixels.*

Figures 9.1 and 10.2 *(bottom)* that crosses three singularity neighbourhoods,
just above two singular points and just below the other. For the experiment,
the error in the initial guess was uniformly 25% of the tuning wavelength
(i.e. $s = 5$ pixels). Figure 11.2 *(top)* shows the results of $\tilde{d}_0(x)$ without the
removal of singularity neighbourhoods. The middle plot shows the results
of $\tilde{d}_0(x)$ with disparity measurements removed whenever either of the left or
the right filter responses did not satisfy the stability constraint (10.6) with
$\tau = 1.25$. The effect of removing these singularity neighbourhoods is clear.
Finally, Figure 11.2 *(bottom)* shows the more accurate results of $\tilde{d}_1(x)$, with
the same singularity neighbourhoods removed.

Sanger (1988) avoids some of the errors that occur in singularity neighbourhoods with the use of smoothing applied to raw disparity estimates and
a constraint on the difference in amplitude between the left and right signals.
The constraint on the amplitude difference will catch some of the incorrect
estimates. In particular, when the amplitude derivative is large, there will
often (but not always) be a large difference in amplitude between the left
and right views. The smoothing reduces many of the remaining errors, but
in doing so it sacrifices resolution and the accuracy of other estimates. Also

note that Sanger did not iterate the approach (as in (11.8)), and therefore the use of the peak tuning frequency in (11.3) limits the magnitude of the errors produced. Iteration would push these errors further from the true disparity (this is shown below in Figure 11.7). Jepson and Jenkin (1989) employed a constraint on disparity gradients in terms of spatial phase differences in the individual views. This constraint is similar to that in (10.4) and removes some of the errors caused by phase behaviour in singularity neighbourhoods. But it was found to be unreliable.

Figure 11.3 shows mean disparity error and standard deviation bars for the old and new predictors ($\tilde{d}_0$ and $\tilde{d}_1$) as a function of filter scale (for $4 \leq \lambda \leq 64$). There was no scale change between the left and right views, and the error in the initial guess was again 25% of the tuned wavelengths (i.e. $s(\lambda) = 0.25\lambda$). Relative prediction error was computed as $e(x) = 100.0\,(s(\lambda) - \tilde{d}_j(x))/s(\lambda)$. Singularity neighbourhoods were detected using (10.4) and (10.5) with $\tau_\rho = 1.0$ and $\tau_k = 1.2$. The improved accuracy of $\tilde{d}_1(x)$ is evident. Also note the significant positive bias in $\tilde{d}_0(x)$. Because real images tend to have more power at lower frequencies, the instantaneous frequency $\phi'(x)$ will often be less than the peak tuning frequency. Therefore the order-s term in (11.5) has a positive average value, which causes the observed bias. This bias is absent from the results of $\tilde{d}_1(x)$ since the local frequency is computed.

Figure 11.4 shows the dependence of $\tilde{d}_1(x)$ on the error in the initial guess $s(\lambda)$ with no scale change between the two views. Disparities were between 2% and 44% of the wavelengths to which the filters were tuned (i.e. $s(\lambda) = \alpha\lambda$, for $\alpha = 0.02, ..., 0.44$). The wavelengths used were between 25 and 100 pixels. Singularity neighbourhoods were detected using (10.6) with $\tau = 1.25$, prediction errors were computed as $e(x) = 100.0\,(s(\lambda) - \tilde{d}_1(x))/\lambda$. Although the results are extremely good, the errors increase dramatically for $s(\lambda) > 0.37\lambda$. To explain this, notice that with $\beta = 0.8$ octaves, the stability constraint with $\tau = 1.25$ removes all points with $\phi'(x) > 1.34\,k(\lambda)$. This means that local wavelengths may be as small as $0.75\,\lambda$, and the domain of convergence may be as small as 0.375λ. As the initial disparities $s(\lambda)$ increased beyond 0.37λ, a larger proportion of initial guesses fell within the wrong domain of convergence. When the boundary between the domains of convergence is crossed, the disparity estimates are wrong by about one wavelength (because the phase difference wraps around modulo 2π). For Figure 11.4 only positive disparities were used. Therefore, a histogram of disparity estimates should reveal two peaks, one at $s(\lambda)$, and the other near $s(\lambda) - \lambda$. This explains the positive error bias for $s(\lambda) > 0.37\lambda$ and the rapid increase in the standard deviation. For negative disparities a corresponding negative bias appears.

Figure 11.5 shows the dependence of $\tilde{d}_1(x)$ on scale differences (up to

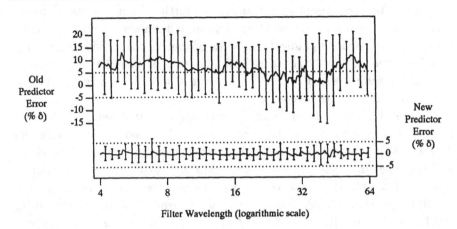

Figure 11.3. Disparity Error for Old and New Predictors: *Mean prediction error and standard deviation bars for $\tilde{d}_0(x)$ and $\tilde{d}_1(x)$ are shown as a function of filter wavelength. The error in the initial guess (i.e. the disparity) was $s = 0.25\lambda$. Disparity error was computed as $e(x) = 100.0\,(s - \tilde{d}_j(x))/s$, $j = 0,1$. Singularity neighbourhoods were removed using (10.4) and (10.5) with $\tau_\rho = 1.0$ and $\tau_k = 1.2$.*

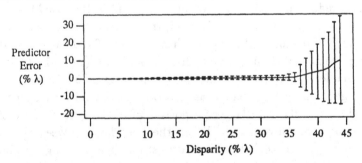

Figure 11.4. Disparity Error versus True Disparity: *Mean disparity error and standard deviation bars are shown as a function of the initial guess error $s(\lambda) = \alpha\lambda$ for $\alpha = 0.02, ..., 0.44$. Disparity error was computed as $e(x) = 100.0\,(s(\lambda) - \tilde{d}_1(x))/\lambda$.*

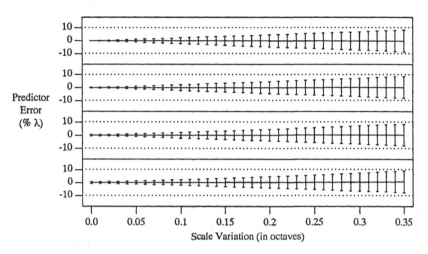

Figure 11.5. Disparity Error versus Scale Variation: *Mean disparity error and standard deviation bars are shown as a function of scale differences (in octaves) between left and right views. Disparity error was computed as $e(x) = 100.0\,(s(\lambda) - \tilde{d}_1(x))/\lambda$. From top to bottom, the four plots correspond to errors in the initial guess of $s(\lambda) = 0.0, 0.1\lambda, 0.2\lambda,$ and 0.3λ.*

35% of an octave).[2] The filter wavelengths used were between 12 and 48 pixels. Scale differences between the two views can be simulated by applying filters with different peak frequencies to a single input signal. The top panel shows prediction error as a function of scale variation between the left and right views when the initial guess was exact; that is, there was scale change between the left and right views locally, but *no* spatial shift. In terms of scale-space expansions (e.g. Figure 9.1), the prediction error here is caused solely by the horizontal drift of level phase contours as scale changes. It is encouraging to note that the prediction errors remain tolerable even for scale changes of 20% of an octave. The next three plots, from top to bottom, show the prediction error as we introduce spatial shifts (errors in the initial guess) of 10%, 20%, and 30% of the tuned wavelength. Again, it is encouraging that the prediction errors increase only slightly. In all cases, the disparity errors remain below 10% of the filter wavelength, despite scale variations of 27.5% (0.35 octaves) between left and right views. These results are consistent with those in Section 9.3 concerning the effect of white Gaussian noise on the expected variation of phase as a function of scale.

---

[2]The relation between relative scale changes and those measured in octaves is straightforward: If λ_1 is α octaves above λ_0, then $\lambda_1 = 2^\alpha \lambda_0$.

Finally, it is important to note that all disparity estimates obtained in singularity neighbourhoods (as defined in Section 10.3) were excluded from the statistics reported in Figures 11.3 – 11.5. The inaccuracies shown in these figures are generally limited to a small fraction of a wavelength, whereas the errors that occur in singularity neighbourhoods are often as large as plus or minus one wavelength. As a consequence, when measurements in singularities neighbourhoods are included the error variances grow dramatically and the error behaviour shown in Figures 11.3 and 11.4 is no longer discernible.

11.4   Application to 2-D Images

The previous examples have shown the effect of this new technique in 1-d. To further demonstrate the results, an existing algorithm [Jepson and Jenkin, 1989] has been modified to incorporate the improved disparity predictor and the removal of singularity neighbourhoods. A coarse-to-fine control strategy based on that of Nishihara (1984) was used. The Gabor filters had bandwidths of one octave, and the different wavelengths to which the filters were tuned were 4, 8, 16, and 32 pixels. Computation begins at the coarsest scale with initial guesses of $d_i = 0$. Computed disparities at one level are then passed down as initial guesses for the next finer level. The results reported below, for both the new and the old methods, are only those disparity estimates computed from the finest channel. The technique is demonstrated on a stereogram of a car part,[3] the left view of which is shown in Figure 11.6. Although not known accurately, the actual disparities range between -1 and 20 pixels.

The goal of this experiment, and Figures 11.7 and 11.8, is to illustrate how important, and how common, are the measurement errors due to phase singularities. As mentioned in Section 11.3 the techniques described in [Sanger, 1988] and [Jepson and Jenkin, 1989] used several conditions and smoothing in an attempt to remove the incorrect estimates. Figure 11.7 shows the output of the old predictor $\tilde{d}_0(x)$ with 1 and then 10 iterations, without the removal of singularity neighbourhoods and without the application of *ad hoc* methods for eliminating incorrect estimates. The occurrences of incorrect measurements are clear. Disparity in both plots is encoded as height. The incorrect estimates are more obvious in the second plot as the disparity estimates have been allowed to converge. With only one iteration, the disparity estimates, $\tilde{d}_0$, are bounded by the model of constant frequency (taken to be the peak tuning frequency). With further iterations the magnitudes of esti-

---

[3]The Renault car part stereogram is a widely available stereo pair that is often used in the literature. The images are 256x256 8-bit images. The car part is roughly 'T' shaped, with the 'T' being slightly inclined in depth.

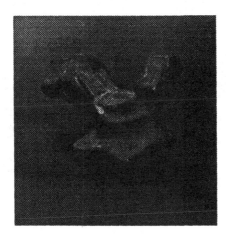

Figure 11.6. **Car Part Stereogram:** *Left image of Renault car part stereo pair.*

mated disparity are not so constrained, and the adverse effects of singularity neighbourhoods become clearer.

Figure 11.8 shows the response of the same algorithm taking advantage of the improved phase-based disparity predictor $\tilde{d}_1(x)$, with only one iteration, after the removal of singularity neighbourhoods. Singularity neighbourhoods were detected using the two stability constraints in (10.4) and (10.5): with this method the results do not change significantly when more iterations are allowed. Figure 11.8 *(top-left)* shows the estimated disparity encoded as intensity. This image clearly shows that the results are well isolated to the car part (areas of black indicate discarded measurements). For illustrative purposes, these estimates were crudely blurred (square-wave averaging over active neighbours with radius of 2 pixels).[4] The result, shown in Figure 11.8 *(top-right)*, better illustrates the depth variation. The same result is also displayed below as a perspective height plot so that comparisons can be made with Figure 11.7. At first glance, the structure found throughout the car part is somewhat difficult to compare with that produced by the old method. However, the noise in the background region around the part has clearly been removed, and upon closer inspection we can see the accuracy and robustness of the new technique, especially at the boundaries of the part.

---

[4]This blurring is used for illustrative purposes only. As discussed in Chapter 5, it is unlikely that the blurring of velocity or disparity measurements, and hence filling in regions without measurements, will lead to more accurate measurements.

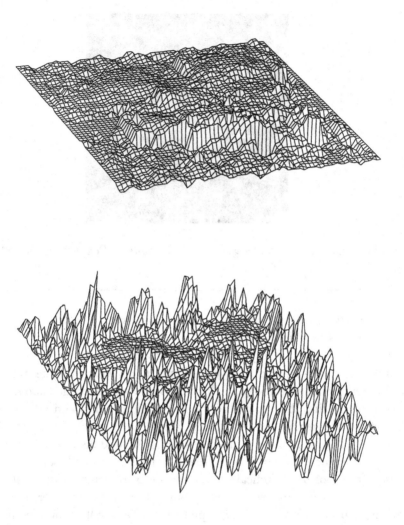

Figure 11.7. Old Predictor Results: *Results of $\tilde{d}_0(x)$ on the car part stereogram without removal of singularity neighbourhoods, after 1 iteration (top) and after 10 iterations (bottom). The results show the disparity estimates obtained at the finest scale of the coarse-to-fine algorithm.*

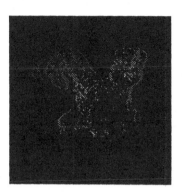
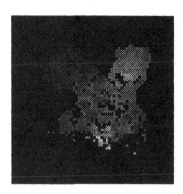

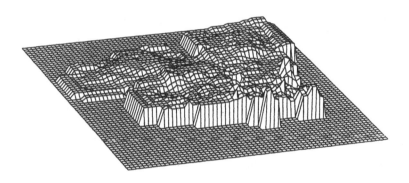

Figure 11.8. **New Predictor Results:** *(top-left) Results of $\tilde{d}_1(x)$ at the finest level of the coarse-to-fine algorithm on car part stereogram after 1 iteration with singularity neighbourhoods removed. (top-right) The results are shown after simple blurring of the disparity estimates over 4×4 neighbourhoods (for illustrative purposes only). (bottom) The blurred results are also shown as a perspective height plot for comparison with Figure 11.7.*

Part IV

Conclusions

Chapter 12

Summary and Discussion

This chapter summarizes the main ideas and results contained in the monograph. It emphasizes the key departures from existing approaches to the problem of measuring image velocity, and it briefly outlines several problems that remain for future research.

12.1 Problem Formulation

In order to measure velocity in image sequences, it is necessary to accurately localize and track image properties in space and time. The goal, as discussed in Chapter 2, is to extract an approximation to the 2-d motion field, a geometric quantity that relates the 3-d locations and velocities of points on objects to 2-d image locations and velocities on the imaging surface. The fundamental challenge is to *define* image velocity and to devise a complementary measurement technique. This is viewed as a problem of defining an image property to be tracked through space-time that evolves according to the 2-d motion field, and thus reflects the motion of surfaces in the actual 3-d scene. A principal difficulty is the fact that image intensity depends on several factors in addition to the camera and scene geometry, such as shading, specular reflection, variations in scene illumination, shadows and transparency.

Chapter 2 developed this perspective using a simple model of time-varying image formation. The model illustrates the geometric deformations that occur as surfaces move in 3-d with respect to the camera, and some of the photometric factors such as shading and specular reflection. With specular reflections, as with shadows under diffuse illumination and transparency, it is reasonable to expect that more than one legitimate velocity may exist within a single local neighbourhood of the image. Therefore, the image property we choose to track should be stable with respect to small geometric deformations, and should be relatively insensitive to contrast variations.

Moreover, it should be possible to simultaneously compute more than one velocity estimate in a single image neighbourhood.

This perspective is an important departure from previous approaches that assume image velocity is unique at every point in the image,[1] and that image translation is a suitable (local) model for the time-varying nature of image intensity. It is argued that these assumptions are overly restrictive, and have led to techniques that are too sensitive to deviations from image translation. It is also argued that current approaches that combine local measurements of component image velocity to determine the full 2-d velocity make the implicit assumption that the different component velocity measurements arise from the same surface in the scene. When this assumption is invalid, these techniques produce incorrect velocity estimates. The approach put forth in this monograph separates the computation of component velocity from the inference of 2-d velocity.

12.2   Measurement of Image Velocity

The measurement technique developed and tested in Part II of the monograph was broken into three main stages of processing: The first involves a decomposition of the input (in space-time) according to a family of velocity-tuned band-pass filters. The second uses the phase behaviour of the output of individual filters to measure component image velocity, and the third is responsible for integrating these local measurements to produce an explicit representation of the 2-d velocity field.

12.2.1   Velocity-Specific Representation

As discussed in Chapter 4, an initial velocity-specific representation of the image sequence has several distinct advantages as a precursor to the computation of image velocity. Most importantly, it helps to separate manifestations of potentially independent scene events that might occur in a single neighbourhood. Because component velocity is measured simultaneously from the output of each velocity-tuned filter, we have the ability to measure the velocity of more than one surface in a single neighbourhood. Other benefits of the prefiltering include noise reduction, and a compact (efficient) encoding of the image sequence.

The design considerations for the family of filters, and the extent to which

---

[1]In Chapter 8 we discussed a method for computing 2-d velocity from component velocity that also makes this assumption. However, this approach was taken for convenience to illustrate the potential accuracy. In general, the integration of local component velocity measurements is beyond the scope of this monograph.

the separation is likely to be successful, were discussed in Chapters 3 and 4. Chapter 3 explained that velocity is a form of orientation in space-time, and derived frequency-domain expressions for the 2-d translation of textured patterns and 1-d profiles. The related issues of spatiotemporal localization (windowing), the uncertainty relation, transparency, occlusion, temporal smoothing/sampling and motion blur were also discussed. Chapter 4 outlined the design and implementation of a family of velocity-tuned filters. Although the representation is not optimal, it is reasonably efficient because the filter tunings do not overlap significantly and the outputs are subsampled. Appendix C describes a method for numerically differentiating the subsampled representation of the outputs of individual filters, since differentiation is needed for the measurement of image velocity.

One drawback of the current implementation is that the numerical differential scheme requires that each filter output be over-sampled at approximately three times the minimal rate. In space-time this implies nine times as many cells working three times as fast. The issue of accurate interpolation and differentiation from the collection of filter outputs sampled at the minimal rate was not addressed and remains a topic for future consideration. Another drawback of the current implementation is the FIR temporal filtering. The implementation discussed in this monograph used a symmetric temporal aperture in time that requires a reasonably long temporal delay and a large amount of temporary memory. It would be desirable to have filters, perhaps recursive, that require less memory and shorter temporal delays while maintaining linear phase behaviour over the effective pass-band of their amplitude spectra. The use of phase information does not depend on the particular form of filters used here.

12.2.2 Component Image Velocity from Local Phase

Chapter 6 defined image velocity in terms of the phase of the output of individual velocity-tuned filters. Component image velocity was defined as the motion normal to level phase contours, measured by the spatiotemporal phase gradient. It was argued that phase information is more stable than the amplitude of the filter output both with respect to contrast variations, such as those caused by changes in illumination or surface orientation, and with respect to changes in scale, speed and image orientation, such as those caused by perspective projection of 3-d objects in space-time. Because of this it is expected to yield accurate velocity estimates both for pure image translation and for deviations from image translation that are common in projections of 3-d scenes.

Chapter 6 also discussed the use of phase information in relation to other approaches. For example, it is much like gradient-based techniques, although here it is applied to the phase signal rather than to the original image intensity or to filter outputs directly. Phase information may also be viewed as a generalization of zero-crossings. But because arbitrary values of phase are used, and not just zeros, a denser set of estimates is usually obtained. Furthermore, the detection and localization of level crossings is not required as it is with zero-crossing tracking. Finally, the phase gradient provides a measure of the instantaneous frequency of the filter output, and is consistent with the definition of 2-d velocity in terms of spatiotemporal frequencies.

The technique's accuracy, robustness, and localization in space-time were demonstrated in Chapter 7 through a series of experiments involving image sequences of textured surfaces moving in 3-d under perspective projection. Several of these involved sizable time-varying perspective deformation. The width of support (in all but one experiment) was limited to 5 pixels (frames) in space (time, respectively) at one standard deviation. The accuracy is typically an order of magnitude better than the initial tuning of the filters. In cases dominated by image translation the results are usually better than this, but with substantial amounts of geometric deformation (relative to the size of the filter support) the results begin to degrade. Successful results were also shown for a simulated case of transparency.

There are, however, several issues that require further examination. One is occlusion. The technique, as it stands, gives incorrect estimates at occlusion boundaries when the background and foreground are highly textured. One reason for this, as discussed in Chapter 3 is that occlusion and nonlinear transparent phenomena can be shown to spread power throughout large regions of frequency space, into channels tuned to very different velocities. The accuracy of velocity estimates also deteriorates as the amount of geometric deformation increases within the support width of the filters. It would be useful to develop an explicit confidence measure, in addition to the constraints listed in Sections 7.2, 10.3, and 11.1, so that unreliable measurements could be detected.

12.2.3   2-D Velocity from Component Velocity

Chapter 8 demonstrated the accuracy with which 2-d velocity might be computed from the component velocity measurements using a least-squares fit of the component velocity estimates in each local neighbourhood to a linear model of 2-d velocity. The resultant 2-d velocity estimates fall mostly within $2°$ of the correct 2-d velocity, which corresponds to relative errors of 6% – 10% (see Figure 8.1). This sub-pixel accuracy compares favourably with current approaches in the literature that report errors between 10% and 30%. The robustness and accuracy of the technique are also evident from the flow fields produced from both synthetic and real image sequences, such as the

Hamburg Taxi Sequence (Figures 8.7 – 8.9).

As mentioned throughout the monograph, we view the integration of local measurements to determine the full 2-d velocity as an inferential process, for which it should not be assumed *a priori* that all estimates arise from the same surface in the scene. Nor should vast amounts of smoothing be applied to fill in regions without measurements or alleviate the problems of noise. The method used in Chapter 8 does not require a subsequent smoothing process to reduce the appearance of noise in the measurements, other than that implicit in the least-squares estimation (cf. [Yuille and Grzywacz, 1988; Nagel and Enkelmann, 1986]). However, it does not take several of the common causes of intensity variations into account, such as occlusion, shadows, specular reflection, and transparency. The integration of basic measurements under more general conditions remains an important topic for further research.

12.3 Phase Properties of Band-Pass Signals

Part III of the monograph dealt specifically with the phase behaviour of band-pass signals in an attempt to gain more insight into the robust use of local phase information. The results help to explain the success of our phase-gradient approach for measuring image velocity, as well as the success of phase-difference techniques for measuring binocular disparity. They are also relevant to zero-crossing and phase correlation techniques.

Chapter 9 examined the stability of phase information with respect to scale perturbations of the input; in a scale-space framework, input scale perturbations were simulated by varying the scale to which the filter is tuned. It was shown that phase is typically stable under small scale perturbations, and it is quasi-linear as a function of spatial position. Both the stability of phase through scale, and its linearity through space were shown to depend on the correlation that exists between filters tuned to nearby scales at nearby spatial locations, and are therefore a function of the form of the filters. For a given filter type, as the bandwidth increases phase becomes stable with respect to larger scale perturbations, but the extent over which phase is expected to be linear decreases. In the context of disparity measurement for example, the bandwidth of the filters should depend, in part, on the magnitudes of deformation that are expected between the left and right views.

Although phase is typically quite stable with respect to small geometric deformations, it was clear from demonstrations in Chapters 6 and 9 that in certain regions phase is sometimes very unstable, and therefore leads to unreliable measurements. Chapter 10 showed that a major cause of phase instability is the occurrence of phase singularities in scale-space. It described a qualitative model of the filter output that generates singularities as the scale of the input changes, which provides a clear picture of the behaviour of band-pass signals in the neighbourhoods about phase singularities within which phase is sensitive to changes in scale and spatial position. This helps to explain the frequent occurrence of singularities, and it serves as the basis for an efficient method of detecting singularity neighbourhoods using constraints on the instantaneous frequency and the relative amplitude derivative of the filter responses. These constraints are crucial for the robustness of phase-based measurement techniques as they help to ensure that poor measurements are eliminated. This analysis also explains a significant source of error in gradient-based techniques and zero-crossing approaches, namely, the instability of level crossings in the output of scale-specific filters.

Finally, Chapter 11 demonstrated the importance of the results in Chapters 9 and 10 using real images in the context of measuring binocular disparity. An improved phase-difference method was described that exploits the instantaneous frequency of the filter response. The results of this disparity predictor, before and after the removal of singularity neighbourhoods were shown for a wide range of scale changes and spatial disparities between left and right views of real images. With one iteration of the predictor, disparities are reliably estimated to within 10% of a wavelength, despite 20% scale changes.

Part III is a first step toward a more thorough understanding of the phase behaviour of band-pass signals. Most of the analysis focused on white noise inputs in one dimension. Further analysis of phase stability and phase singularities is required with a more realistic model of local image structure in two dimensions. In addition, greater attention should be given to the accuracy of velocity measurement in the face of photometric distortions and multiple local velocities. Despite this work that remains, Part III constitutes important progress towards a sound understanding of phase-based techniques for the measurement of image velocity and binocular disparity.

Appendix A

Reflectance Model

In modelling the expected forms of spatiotemporal intensity variations, we exploit a relatively simple scene model incorporating smooth surfaces, with both diffuse and specular components of reflection. The local geometric deformation that depends on the camera and scene geometry is discussed in Section 2.2.1. This appendix concentrates on the photometric factors, that is, the surface reflectance function and the principal sources of illumination. Rather than attempt to provide a detailed physical model, we have tried to sketch a simple model and the consequent sources of intensity variation. This helps to demonstrate the influence of shading and specular reflection, the importance of which would not be lessened in a more realistic model.

The surface reflectance model that we adopt consists of a diffuse (Lambertian) term and a specular term. Such models have been used in computer graphics with reasonable success to render realistic images [Phong, 1975; Horn, 1977].

Photometry

In what follows we assume a single smooth surface, and let $\mathbf{p}$ denote a point on the surface in a local surface-based coordinate system. Let $L(\mathbf{p}, t)$ denote the scene radiance (in watts per square meter per steradian) in the direction of the camera. Following Horn (1986), image irradiance (in watts per square meter) is proportional to scene radiance:

$$E_I(\mathbf{x}, t) \;=\; L(\mathbf{p}, t)\,\frac{\pi}{4}\left(\frac{d_l}{f}\right)^2 \cos^4\alpha\,, \qquad (A.1)$$

where E_I denotes image irradiance, d_l is the diameter of the lens aperture, and α denotes the angle between the optical axis and the viewing direction towards the surface (see Figure A.1). The factors that govern scene radiance include the *bidirectional reflectance function* of the surface $R(\mathbf{d}_i, \mathbf{d}_e; \mathbf{p})$ (the

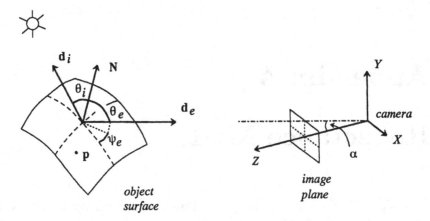

Figure A.1. Surface Reflectance: *Surface reflectance properties are given in terms of a surface-based coordinate system, where surface locations are given by* **p**. *Then, at a given surface point, the directions of incidence (toward illuminating sources) and emittance (toward the camera) are specified in spherical coordinates (θ, ψ), where θ denotes the polar angle, and ψ denotes a direction on the surface with respect to some reference line.*

ratio of reflected radiance to incident irradiance as a function of the incident and emittant directions), and the directions of the illumination sources and the camera with respect to the surface normal. These directions are conveniently specified in surface-based, spherical coordinates $\mathbf{d} \equiv (\theta, \psi)^T$ (see Figure A.1).

Let the surface irradiance in terms of foreshortened surface area at point **p** from direction $\mathbf{d}_i$ be denoted by $E_S(\mathbf{d}_i; \mathbf{p}) \equiv E_L(\mathbf{d}_i; \mathbf{p}) \cos \theta_i$, where E_L denotes the surface irradiance in watts per square meter normal to $\mathbf{d}_i$. The total irradiance at **p** is given by the integral of E_S over the hemisphere of possible incident angles:

$$\bar{E}_S(\mathbf{p}) = \int_{-\pi}^{\pi} \int_0^{\pi/2} E_S(\mathbf{d}_i; \mathbf{p}) \sin \theta_i \, d\theta_i \, d\psi_i . \qquad (A.2)$$

Scene radiance can be written as the integral of surface irradiance multiplied by the reflectance function, with the emittance angle equal to the direction of the focal point of the camera:

$$L(\mathbf{d}_e; \mathbf{p}) = \int_{-\pi}^{\pi} \int_0^{\pi/2} R(\mathbf{d}_i, \mathbf{d}_e; \mathbf{p}) \, E_S(\mathbf{d}_i; \mathbf{p}) \sin \theta_i \, d\theta_i \, d\psi_i . \qquad (A.3)$$

In the model adopted here, the reflectance function is written as the sum of a diffuse term and a specular term. The surface radiance arising from the diffuse (Lambertian) component is independent of the camera direction and can be written as a surface-dependent albedo function $a(\mathbf{p})$ multiplied by the total surface irradiance $\bar{E}_S$. The specular term depends strongly on the viewing direction and surface roughness. Our crude model assumes that the specularly reflected light, given irradiance direction $\mathbf{d}_i$, decreases as a function of the angle γ between the direction of emittance $\mathbf{d}_e$, and the direction of perfect specular reflection $\mathbf{d}_i + (0, \pi)^T$. Following Horn (1977) and Phong (1975), the reflectance function may be simplified to

$$R(\mathbf{d}_i; \mathbf{d}_e; \mathbf{p}) \;=\; (1 - \lambda)\, a(\mathbf{p}) \;+\; \lambda\, \frac{n+1}{2}\, \cos^n \gamma\,, \qquad (A.4)$$

where $\lambda \in (0, 1)$ determines the relative strength of the specularity, and $\cos \gamma \equiv (2 \cos \theta_i \cos \theta_e - \cos \theta_p)$, where θ_p is the angle between directions $\mathbf{d}_i$ and $\mathbf{d}_e$ [Horn, 1977]. From (A.2), the scene radiance (A.3) simplifies to

$$L(\mathbf{d}_e; \mathbf{p}) \;=\; (1 - \lambda)\, a(\mathbf{p})\, \bar{E}_S(\mathbf{p}) \;+$$
$$\lambda\, \frac{n+1}{2} \int_{-\pi}^{\pi} \int_0^{\pi/2} \cos^n \gamma\, E_S(\mathbf{d}_i; \mathbf{p})\, \sin \theta_i\, d\theta_i\, d\psi_i\,. \qquad (A.5)$$

It is often convenient to separate the illumination due to the principal light source (modelled as a distant point source), from the illumination due to other, secondary or ambient sources. If we ignore the effects of mutual illumination on the diffuse term, and take into account only the direction of a principal light source $\mathbf{L}$ in relation to the surface normal $\mathbf{N}(\mathbf{p})$, where both $\mathbf{L}$ and $\mathbf{N}(\mathbf{p})$ have unit length, then $\bar{E}_S(\mathbf{p})$ simplifies to $E_L(\mathbf{p})\,\mathbf{N}^T\mathbf{L}$. Under the same conditions the specular reflection due to the principal light source simplifies to $E_L(\mathbf{p})\,(\mathbf{D}^T\mathbf{R})^n$ where $\mathbf{D}$ is the unit vector in the direction of the camera from $\mathbf{p}$, and $\mathbf{R}$ is the direction of perfectly specular reflection: $\mathbf{R} = -[\,\mathbf{I} - 2\,\mathbf{N}\mathbf{N}^T\,]\,\mathbf{L}$. In this case, (A.5) simplifies to

$$L(\mathbf{d}_e; \mathbf{p}) \;=\; E_L(\mathbf{p}) \left[(1 - \lambda)\, a(\mathbf{p})\,\mathbf{N}^T\mathbf{L} \;+\; \lambda\, \frac{n+1}{2}\, (\mathbf{D}^T\mathbf{R})^n \right]. \qquad (A.6)$$

The specular reflections of principal light sources are referred to as highlights, and are commonly thought to be the main consequence of the specular component of surface reflection. However, because the power of specular reflections often exceeds that of diffuse reflection, other effects due to secondary sources (mirror-like reflection of neighbouring objects) should also be considered.

Time-Varying Image Behaviour

We now combine the geometric and photometric factors to obtain a model of temporal intensity variations caused by the relative motion of the surface with respect to the camera. Image intensity is assumed to be equal to image irradiance, and hence proportional to scene radiance in the direction of the camera (A.1). Also, because of the superposition of the diffuse and specular components of surface reflectance (A.4), we deal with their time-varying behaviour separately.

Diffuse Component

First, consider the diffuse term. At some time t, the image of a surface patch with coordinates $p(x, t)$ is given by

$$
\begin{aligned}
I(x(p, t), t) &= c\, a(p)\, \bar{E}_S(p, t) \\
&= S(p, t)\, I(x(p, 0), 0) ,
\end{aligned} \tag{A.7}
$$

where

$$
S(p, t) = \frac{\bar{E}_S(p, t)}{\bar{E}_S(p, 0)} \tag{A.8}
$$

denotes a shading function, and c incorporates constants from (A.1) and (A.5).[1] If we consider a single, distant light source, then $S(p, t)$ becomes[2]

$$
S(p, t) = \frac{N(p, t)^T L(t)}{N(p, 0)^T L(0)} . \tag{A.9}
$$

It is instructive to express $S(p, t)$ to first-order about $p = 0$ and $t = 0$, as

$$
S(p, t) = 1 + t\, S_t|_{p=0, t=0} + p^T S_p|_{p=0, t=0} + O(\| (p, t) \|^2). \tag{A.10}
$$

The second term, $t\, S_t$, represents a (spatially) homogeneous scaling of intensity (contrast variation). It is due primarily to a change in angle between the principal light sources and the surface normal at $p = 0$. The next term, $p^T S_p$, represents a shading gradient, which is primarily a function of surface curvature, and illumination due to nearby light sources.

Finally, using (2.7) and (A.7), we can express the image at time t as a local geometric deformation and reshaded version of the image at time $t = 0$; that is,

$$
I(x(p, t), t) = S(p, t)\, I(x(p, 0), 0) , \tag{A.11}
$$

---

[1] For simplicity, this assumes that $a(p)$ is a function of surface position only, and does not change with time. It also ignores temporal variations in the angle between the optical axis and the visual direction of p (α in (A.1)). With these assumptions, temporal variations in image intensity are due solely to geometric deformation and changes in surface irradiance.

[2] Although, in principle $E(p)$ could be a function of time, here we assume for simplicity that it is constant.

where

$$\mathbf{x}(\mathbf{p},\, t) \;\approx\; \mathbf{x}_0(t) + B\,[\mathbf{x}(\mathbf{p},0) - \mathbf{x}_0(0)]\;,$$

$\mathbf{x}_0(t) = \mathbf{x}_0(0) + t\,\mathbf{v}(\mathbf{x}_0(0),\, 0)$, and $B = \mathrm{I} + t\,A$. In combination, (A.10) and (A.11) yield a crude model of the time-varying intensity behaviour for diffuse surfaces in terms of geometric deformation and smooth contrast variation. The model implicitly makes a sort of general view-point assumption that points stay visible through space-time, and that the local shading function changes smoothly. That is, self-occlusion and lighting conditions should not combine to cause large geometric deformations or dramatic variations in the total incident illumination $\bar{E}_S$ over short time intervals.

Specular Component

In order to consider the time-varying behaviour of images due to specular reflection we concentrate on the ideal case of a smooth (locally planar) surface with perfectly specular (mirror) reflection (see Figure A.2). Let $\mathbf{M}_0$ be a point on the mirror with normal $\mathbf{N}$ in viewer-centred coordinates. An arbitrary point $\mathbf{X}_0$ on another surface (on the appropriate side of the mirror) can be thought of as a light source and will be reflected *through* the mirror to a *virtual* point $\mathbf{Y}_0$ that satisfies

$$\mathbf{Y}_0 \;=\; [\,\mathrm{I} - 2\mathbf{N}\mathbf{N}^T\,](\mathbf{X}_0 - \mathbf{M}_0) + \mathbf{M}_0\,. \tag{A.12}$$

The 3-d velocity of the virtual point $d\mathbf{Y}_0/dt$ is a function of the 3-d velocities of $\mathbf{M}_0$ and $\mathbf{X}_0$, as well as the temporal variation in $\mathbf{N}$. In general, the instantaneous motion of the mirror is independent of the motion of the reflected surface that contains $\mathbf{X}_0$. However, there are two special cases worth mentioning: *1)* the motion of $\mathbf{Y}_0$ resulting from motion of $\mathbf{X}_0$ while the camera remains stationary with respect to the mirror; and *2)* the induced motion of $\mathbf{Y}_0$ resulting from camera motion only.

In the first case, where the camera and mirror are effectively stationary, the motion of the virtual point satisfies

$$\frac{d\mathbf{Y}_0}{dt} \;=\; [\,\mathrm{I} - 2\mathbf{N}\mathbf{N}^T\,]\frac{d\mathbf{X}_0}{dt}\,, \tag{A.13}$$

and is depicted in Figure A.2 *(left)*. In this case the 3-d velocity of $\mathbf{X}_0$ is simply reflected through the mirror using the same projection as the static case. The mirror creates a virtual surface that moves independently of the camera, but is constrained by $d\mathbf{X}_0/dt$ and $\mathbf{N}$.

The second case, where camera moves while the mirror and the surface point are assumed to be stationary, is depicted in Figure A.2 *(right)*. In the special case of a planar mirror, it can be shown that the motion of

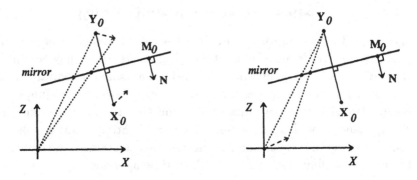

Figure A.2. Mirror Geometry: *Given a planar mirror containing* M_0, *with normal* N, *points* X_0 *are reflected through the mirror to virtual points* Y_0. *(left) When the camera remains stationary, the velocity of* Y_0 *is equal to the velocity of* X_0 *reflected through the mirror (indicated by dashed vectors). (right) Motion of the camera induces motion of* Y_0. *In both cases, the motion field of the virtual point is different from that of the mirror (indicated by the movement of the intersection of the dotted lines with the mirror).*

Y_0 satisfies the same rigid-body motion parameters as the motion of X_0. More precisely, because only the camera moves, the relative instantaneous translation and rotation of M_0 and X_0 are identical. Let them be T_0 and Ω_0. The instantaneous motion of N is $dN/dt = -\Omega_0 \times M_0$; translation of the mirror with respect to the camera does not affect the direction of its normal. Then, after differentiating (A.12) and some algebraic manipulation, one can show that

$$\frac{d\,Y_0}{dt} \;=\; -T_0 \,-\, \Omega_0 \times Y_0. \qquad (A.14)$$

This shows that the actual reflecting surface (the mirror) and the virtual points (surfaces) undergo the same rigid-body motion. In principle, measurements of the motion field induced by virtual surfaces could be used to measure egomotion.

Finally, in all cases, note that the 2-d motion field associated with virtual surfaces will be different from the motion field of the mirror. As either the camera or X_0 moves, as shown in Figure A.2, the relative location of the virtual point effectively slides along the mirror. Alternatively, the depths of the mirror and virtual point are different, and therefore their respective 2-d motion fields will also be different.

In practice we rarely encounter perfectly specular surfaces. Most surfaces have a combination of specular and diffuse reflection. Surface radiance L typically depends on both the albedo variation of a diffuse component and the directional surface radiance of a specular component. It is also uncommon to encounter a perfectly smooth surface for which the specularly reflecting surface microfacets are all aligned with the surface normal [Cook and Torrance, 1982]. In crude terms (as in (A.4) following [Horn, 1977]), as the distribution of surface microfacets about the surface normal increases, or as surface roughness increases, the reflected pattern of surface irradiance becomes more blurred.

Upon casual inspection of the world, several other consequences of specular reflection are evident and worth mentioning. First, when the reflected surface is distant (far from the reflecting surface) the specular component of surface radiance (reflection) is sensitive to changes in the viewing direction and surface orientation. The result is that motion of the reflecting surface can induce extremely fast 2-d velocities in the motion field of the virtual surface; the change in scene location projected to the camera, given a small change in surface orientation (or mirror), is proportional to the distance of the virtual point from the reflecting surface. This also means that slight, even smooth, variations in the surface normal across the reflecting surface can cause significant non-rigid distortion of the projected pattern. Secondly, because of the significant power of the specular reflection in comparison to the diffuse component, and its sensitivity to changes in surface reflectance properties and perturbations of the surface normal, specular reflection often tends to enhance the visibility properties of the actual (physical) reflecting surface. This can be seen, for example, in the highlights off a polished tile floor, where cracks between the tiles involve variations in the surface normal as well as changes in reflectance properties when, for example, they are filled with dirt.

Our model of image irradiance is rather crude,[3] and yet, with respect to the measurement of visual motion, it is complex enough to illustrate several factors that govern spatiotemporal variations in image intensity. In summary, the model assumes the superposition of diffuse and specular components of reflection. The diffuse component yields spatiotemporal patterns of scene radiance that reflect the properties of the actual radiating (reflecting) surface, including geometric deformation and shading variations. The specular

---

[3]Many factors have not been discussed in this model: No mention has been made of the sensors, which should include gain control, and low-pass spatiotemporal smoothing in order to avoid severe aliasing upon discrete sampling. In discussing surface radiance with respect to (A.1) and (A.3), we have ignored the possibility of photon transmission through the radiating surface. The analysis did not deal with the motion of shadows under diffuse conditions, complex self-shadowing, or anisotropic reflectance properties. Nor has it dealt with occlusion.

component produces patterns of scene radiance that reflect the behaviour of a degraded image of a virtual surface. They can be reasonably clear renditions of virtual surfaces in the case of near-perfect mirrors (e.g. reflections in windows and puddles), but are generally distorted (blurred) and somewhat non-rigid because of variation of the camera position or the surface normal (e.g. consider the back of a spoon). Finally, it was also pointed out that the specular component, especially in the case of highlights, often enhances the visibility of features inherent in the actual reflecting surface such as changes in reflectance or small physical surface markings.

Appendix B

Proof of an n-D Uncertainty Relation

This appendix proves a general n-dimensional expression of the uncertainty relation. Given a signal $f(\mathbf{x})$, with $\mathbf{x} \in \mathbb{R}^n$ and Fourier transform $\hat{f}(\mathbf{k})$, let C and $\hat{C}$ denote the covariance matrices corresponding to the distributions $|f(\mathbf{x})|^2$ and $|\hat{f}(\mathbf{k})|^2$ (assumed to be mean zero). That is, the ij-elements of C and $\hat{C}$, written as c_{ij} and $\hat{c}_{ij}$, are the second-order moments given by

$$c_{ij} = \frac{\int ... \int x_i\, x_j\, |f(\mathbf{x})|^2\, dx_1...dx_n}{\int ... \int |f(\mathbf{x})|^2\, dx_1...dx_n} \,, \qquad (\text{B.1a})$$

$$\hat{c}_{ij} = \frac{\int ... \int k_i\, k_j\, |\hat{f}(\mathbf{k})|^2\, dk_1...dk_n}{\int ... \int |\hat{f}(\mathbf{k})|^2\, dk_1...dk_n} \,. \qquad (\text{B.1b})$$

The variances along coordinate axes are given by $\sigma_j^2 = c_{jj}$ and $\hat{\sigma}_j^2 = \hat{c}_{jj}$ The n-dimensional uncertainty relation that is proved below is given by

$$|C|^{\frac{1}{2}}\, |\hat{C}|^{\frac{1}{2}} \;\geq\; \frac{1}{2^n} \,, \qquad (\text{B.2})$$

where $|C|$ denotes the determinant of C.

The proof for (B.2) follows from a 1-d result in Papoulis (1968) that specifies

$$\sigma_j\, \hat{\sigma}_j \;\geq\; \frac{1}{2} \,. \qquad (\text{B.3})$$

To begin, note that we are free to rotate the coordinate system without affecting the validity of (B.3). Equation (B.3) can therefore be expressed as

$$\sigma_\mathbf{u}\, \hat{\sigma}_\mathbf{u} \;\geq\; \frac{1}{2} \,, \qquad (\text{B.4})$$

where, with $\| \mathbf{u} \| = 1$,

$$\sigma_{\mathbf{u}}^2 = \mathbf{u}^T C \mathbf{u},$$

(B.5a)

$$\hat{\sigma}_{\mathbf{u}}^2 = \mathbf{u}^T \hat{C} \mathbf{u}.$$

(B.5b)

Raise both sides of (B.4) to the power n, and multiply by $\gamma(\vec{\theta}) \equiv |J| r^{1-n}$, where J is the Jacobian of the change of variables from n-dimensional Euclidean coordinates to spherical coordinates $(r, \theta_1, ..., \theta_{n-1})$ [Kendall, 1961]. This yields

$$\sigma_{\mathbf{u}}^n \, \hat{\sigma}_{\mathbf{u}}^n \, \gamma(\vec{\theta}) \geq \left(\frac{1}{2}\right)^n \gamma(\vec{\theta}).$$

(B.6)

Then, with the direction of $\mathbf{u}$ specified by angles $\theta_1...\theta_{n-1}$, we can integrate over the angles in n-d, since (B.4) holds for all directions $\mathbf{u}$, to get

$$\int ... \int \sigma_{\mathbf{u}}^n \, \hat{\sigma}_{\mathbf{u}}^n \, \gamma(\vec{\theta}) \, d\vec{\theta} \geq \int ... \int \frac{1}{2^n} \gamma(\vec{\theta}) \, d\vec{\theta}.$$

(B.7)

Equation (B.7) can be rewritten using the Schwartz inequality in the form $\int f^2 \int g^2 \geq (\int fg)^2$ as follows:

$$\left(\int ... \int \sigma_{\mathbf{u}}^{2n} \, \gamma(\vec{\theta}) \, d\vec{\theta}\right) \left(\int ... \int \hat{\sigma}_{\mathbf{u}}^{2n} \, \gamma(\vec{\theta}) \, d\vec{\theta}\right) \geq \left(\int ... \int \frac{1}{2^n} \gamma(\vec{\theta}) \, d\vec{\theta}\right)^2$$

(B.8)

This can be rewritten as

$$\left(\int ... \int \int_0^{\sigma_{\mathbf{u}}^2} r^{n-1} \, \gamma(\vec{\theta}) \, dr \, d\vec{\theta}\right) \left(\int ... \int \int_0^{\hat{\sigma}_{\mathbf{u}}^2} r^{n-1} \, \gamma(\vec{\theta}) \, dr \, d\vec{\theta}\right)$$

$$\geq \left(\int ... \int \int_0^{1/2} r^{n-1} \, \gamma(\vec{\theta}) \, dr \, d\vec{\theta}\right)^2$$

(B.9)

where $r^{n-1} \gamma(\vec{\theta}) = |J|$ is the determinant of the Jacobian of the change from Euclidean to spherical coordinates.

The integrals in (B.9) represent the volumes of n-d regions, the radii for which are functions of the unit direction vector $\mathbf{u}$. In particular, the two on the left side are hyperellipses with radii $\sigma_{\mathbf{u}}^2$ and $\hat{\sigma}_{\mathbf{u}}^2$ (B.5). The right side amounts to the square of the volume of a hypersphere of radius $\frac{1}{2}$. Finally, it can be shown that (B.9) simplifies to [Kendall, 1961]

$$\left(|C| \frac{2\pi^{\frac{n}{2}}}{n\,\Gamma(\frac{n}{2})}\right) \left(|\hat{C}| \frac{2\pi^{\frac{n}{2}}}{n\,\Gamma(\frac{n}{2})}\right) \geq \left(\frac{1}{2^n} \frac{2\pi^{\frac{n}{2}}}{n\,\Gamma(\frac{n}{2})}\right)^2.$$

(B.10)

The uncertainty relation (B.2) follows directly from (B.10).

Appendix C

Numerical Interpolation and Differentiation of $R(\mathbf{x}, t)$

This appendix outlines a method for the direct numerical interpolation and differentiation of the filter output based on convolution with a discrete kernel. For notational convenience here, let the space-time variables $(\mathbf{x}, t)^T$ be denoted by $\mathbf{x} = (x_1, x_2, x_3)^T$, and their respective Fourier variables $(\mathbf{k}, \omega)^T$ by $\mathbf{k} = (k_1, k_2, k_3)^T$. Also, let the space-time sampling distances be denoted by $\mathbf{S} = (S_1, S_2, S_3)^T$, and let the nodes of the sampling lattice (i.e. the sampling locations) be given by $S[\mathbf{m}] = \sum_{j=1}^{3} m_j S_j \mathbf{e}_j$, where $\mathbf{m} \in \mathbb{Z}^3$ and the vectors $\mathbf{e}_j$ are the standard basis vectors for $\mathbb{R}^3$.

In deriving an interpolant for $R(S[\mathbf{m}])$ it is convenient to view the band-pass response as

$$R(\mathbf{x}) \;=\; M(\mathbf{x})\, C(\mathbf{x}) \,, \tag{C.1}$$

where $C(\mathbf{x}) = e^{i\mathbf{x}^T \mathbf{k}_0}$, and $\mathbf{k}_0$ is the peak tuning frequency of the filter in question. Note that, because $R(\mathbf{x})$ is band-pass with its power concentrated near $\mathbf{k}_0$, $M(\mathbf{x})$ will be low-pass. (This follows from the modulation property (1.9).) From (C.1), $\nabla R(\mathbf{x}, t)$ has the form

$$
\begin{aligned}
\nabla R(\mathbf{x}) \;&=\; \nabla M(\mathbf{x})\, C(\mathbf{x}) \;+\; M(\mathbf{x})\, \nabla C(\mathbf{x}) \\
&=\; \nabla M(\mathbf{x})\, C(\mathbf{x}) \;+\; i\mathbf{k}_0 R(\mathbf{x}) \,.
\end{aligned}
\tag{C.2}
$$

Because $M(\mathbf{x})$ is a low-pass signal it may be interpolated and differentiated from a subsampled representation using standard methods [Dudgeon and Mersereau, 1984]. This is treated below as a precursor to the explicit interpolation and differentiation of $R(S[\mathbf{m}])$.

A subsampled encoding of $M(\mathbf{x})$ (i.e. $M(S[\mathbf{m}]) = R(S[\mathbf{m}])\, C(-S[\mathbf{m}])$) can be interpolated as

$$\tilde{M}(\mathbf{x}) \;=\; \sum_{m_1, m_2, m_3} M(S[\mathbf{m}])\, Q(\mathbf{x} - S[\mathbf{m}]) \,, \tag{C.3}$$

where $Q(\mathbf{x})$ is an appropriate interpolation kernel. If $M(\mathbf{x})$ were perfectly low-pass, then the appropriate interpolant $Q(\mathbf{x})$ would be a (separable) product of three sinc functions [Dudgeon and Mersereau, 1984]. The derivatives of $\tilde{M}(\mathbf{x})$ have the same form:

$$\tilde{M}_{x_j}(\mathbf{x}) \;=\; \sum_{m_1, m_2, m_3} M(S[\mathbf{m}]) \, Q_{x_j}(\mathbf{x} - S[\mathbf{m}]) \,. \qquad (C.4)$$

Therefore both interpolation and differentiation in (C.3) and (C.4) can be viewed as cascades of three 1-d convolutions. If $\nabla\phi(\mathbf{x})$ is computed only at nodes of the sampling lattice, then $\tilde{M}(\mathbf{x})$ is given directly by the subsampled representation (without interpolation), and (C.4) reduces to a single 1-d convolution with the differentiated interpolant.

To obtain expressions for $\tilde{R}(\mathbf{x})$ and $\nabla\tilde{R}(\mathbf{x})$ in terms of explicit interpolation and differentiation of the subsampled filter output $R(S[\mathbf{m}])$, we can replace $M(S[\mathbf{m}])$ in (C.3) and (C.4) with $R(S[\mathbf{m}]) \, C(-S[\mathbf{m}])$. For instance, (C.4) becomes

$$\tilde{M}_{x_j}(\mathbf{x}) \;=\; C(-\mathbf{x}) \sum_{m_1, m_2, m_3} R(S[\mathbf{m}]) \left[C(\mathbf{x} - S[\mathbf{m}]) \, Q_{x_j}(\mathbf{x} - S[\mathbf{m}]) \right] . \,(C.5)$$

Instead of demodulating the filter output before subsampling, we simply modulate the appropriate interpolation/differentiation kernel. After substitution into (C.1) and (C.2) we now have expressions for $R(\mathbf{x}, t)$ and $\nabla R(\mathbf{x}, t)$ in terms of the subsampled filter output.

In the experiments reported in Chapter 7, $\nabla\phi(\mathbf{x})$ was computed only at the nodes of the sampling lattice (i.e. with $\mathbf{x} = S[\mathbf{m}]$ for some $\mathbf{m} \in \mathbb{Z}^3$). In this case, $R(\mathbf{x})$ is given explicitly at the sampling points, and its derivative reduces to

$$
\begin{aligned}
\tilde{R}_{x_j}(\mathbf{x}) &= \tilde{M}_{x_j}(\mathbf{x}) \, C(\mathbf{x}) \;+\; i(\mathbf{e}_j^T \mathbf{k}_0) \, R(\mathbf{x}) \\
&= C(\mathbf{x}) \sum_n M(\mathbf{x} - n\mathbf{e}_j) \, h(n) \;+\; i(\mathbf{e}_j^T \mathbf{k}_0) R(\mathbf{x}) \\
&= C(\mathbf{x}) C(-\mathbf{x}) \sum_n R(\mathbf{x} - n\mathbf{e}_j) \left[e^{in(\mathbf{e}_j^T \mathbf{k}_0)} h(n) \right] \;+\; i(\mathbf{e}_j^T \mathbf{k}_0) R(\mathbf{x}) \\
&= \sum_n R(\mathbf{x} - n\mathbf{e}_j) \, H(n) \;+\; i(\mathbf{e}_j^T \mathbf{k}_0) R(\mathbf{x}) \,, \qquad (C.6)
\end{aligned}
$$

where $h(n)$ is an appropriate kernel for numerical differentiation of a low-pass signal, and $H(x) = h(x) \, c(x)$ is the new kernel that is to be applied directly to $R(S[\mathbf{m}])$.

For appropriately band-limited signals, the interpolation error decreases as the spatiotemporal extent of the interpolating kernel $h(x)$ increases. However, it is desirable to limit its extent in the interests of efficiency and localization in space-time. In comparison to a truncated sinc function, more accurate interpolants can be derived using low-order polynomials, splines, or optimization techniques [Hou and Andrews, 1978; Oppenheim and Schafer, 1975; Schafer and Rabiner, 1973]. The choice of interpolant is important as it affects the choice of subsampling rate for the Gabor outputs. In general, an appropriate sampling rate depends on several factors including *1)* the input spectral density, *2)* the form of interpolation, and *3)* a tolerance bound on reconstruction error. For 1-d signals with Gaussian power spectra, interpolation with local polynomial interpolants (e.g. 4 or 5 points) is generally possible to within 5% RMS error if the sampling rate is one complex sample every σ [Gardenhire, 1964; Ty and Venetsanopoulos, 1984]. We used this rate for the implementation discussed in Chapter 7. Because we measure derivatives, it should not be viewed as overly generous. On the other hand, a higher sampling rate means a significant increase in computational expense. Clearly, the relationship between sampling rates and appropriate form of interpolation/differentiation deserves further attention, beyond the scope of the dissertation.

The actual numerical differentiation method used in Chapter 7 to compute $\nabla R(\mathbf{x}, t)$ is based on a standard 5-pt central-difference formula. The coefficients $h(n)$ are taken from the vector $\frac{1}{12s}(-1, 8, 0, -8, 1)^T$, where s is the sampling distance. The corresponding kernel that is applied to $R(S[\mathbf{m}])$, as in (C.6), to find $\tilde{R}_{x_j}(\mathbf{x}, t)$ is therefore given by coefficients $H(n)$ that are taken from the vector $\frac{1}{12s}(-e^{-i2sk_j}, 8e^{-isk_j}, 0, -8e^{isk_j}, e^{i2sk_j})^T$, where $k_j = \mathbf{e}^T \mathbf{k}_0$.

Appendix D

Additional Experiments

The following sections describe experiments that complement those in Chapters 7 and 8 dealing with the measurement of component image velocity, and the application of the entire procedure for measuring 2-d velocity to real image sequences.

D.1 Measurement of Component Velocity

In Sections 7.3 – 7.5 the *tree* image was used exclusively as surface texture so that results from different types of image deformations could be compared easily. Here we give results from similar deformations but with different images. We will refer to the images shown in Figure D.1 as the *car* image and the *house* image. The house image has considerably less structure than the tree image, the car image more. While the number of velocity estimates for experiments with the tree image was normally about 10^4, the typical numbers of estimates in experiments with the car and house images were about 11000 and 5000 respectively. The relevant camera and scene parameters are given in Section 7.3; the images were 150×150 pixels wide.

Side-View Camera Motion

As in Section 7.3 there are three cases of interest: *1)* the surface being perpendicular to the line of sight; *2)* a speed gradient in the direction of image velocity causing dilation and shear; and *3)* a speed gradient normal to the direction of image velocity causing curl and shear.

The first two experiments involved the car image with the surface perpendicular to the line of sight, the results of which are shown in Figure D.2 *(top)*. The velocities in each case were constant throughout the image, with speeds of 0.75 and 2.5 pixels/frame; their scene parameters were ($\alpha_c = 90$; $\alpha_x = \alpha_y = 0$; $d_0 = 15$; and $v_c = 0.075$) and ($\alpha_c = 90$; $\alpha_x = \alpha_y = 0$;

Figure D.1. **Car and House Images:** *Used as texture maps.*

$d_0 = 20$; and $v_c = 0.2$) respectively. Like all the experiments in Section 7.3 the velocity errors increase as distance increases between the estimated local frequency and the corresponding filter tunings. Figure D.3 *(left)* shows error as a function of image location. It is clear from this that the car image contained denser structure at short spatiotemporal wavelengths than the tree image.

The next two experiments used the house image for texture-mapping (see Figure D.2 *(middle)*). Both motions involved speed gradients parallel to the direction of camera motion, so that the induced image speeds varied slightly from the left to the right side of the image with non-zero divergence and shear. The surface gradient in the first case was 15° and image speeds ranged between 0.65 and 0.85 pixels/frame (scene parameters: $\alpha_c = 90$; $\alpha_x = 15$; $\alpha_y = 0$; $d_0 = 15$; $v_c = 0.075$), while in the second case the surface gradient was 25° and speeds ranged between 1.15 and 1.85 pixels/frame (scene parameters: $\alpha_c = 90$; $\alpha_x = 25$; $\alpha_y = 0$; $d_0 = 13$; $v_c = 0.13$). Although the error behaviour is very similar to other experiments, the density of estimates is lower owing to the relatively sparse image structure. This is evident in Figures D.1 *(right)* and in D.3 *(right)* which shows average component velocity error as a function of image location.

Finally, Figure D.2 *(bottom)* shows error histograms from two image sequences in which the speed gradient was vertical while the direction of motion was to the right. The first case (with the car image) involved speeds from 0.875 pixels/frame at the top of the image to 2.125 on the bottom (scene parameters: $\alpha_c = 90$; $\alpha_x = 0$; $\alpha_y = 40$; $d_0 = 11$; $v_c = 0.075$). The second case (with the house image) had speeds of 1.07 at the top and 1.93 at the bottom (scene parameters: $\alpha_c = 90$; $\alpha_x = 0$; $\alpha_y = 30$; $d_0 = 13$; $v_c = 0.13$). The

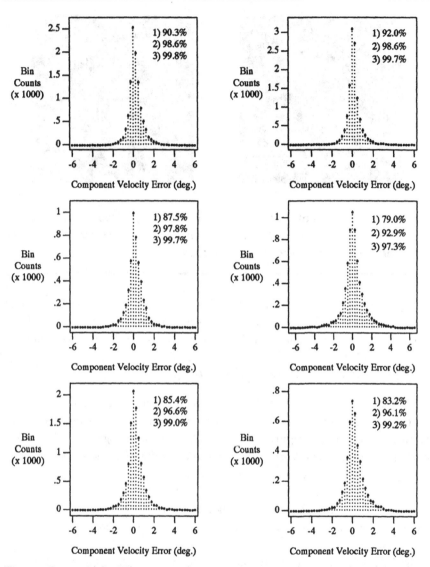

Figure D.2. **Side-View Motion:** *Histograms of velocity errors (7.2), and proportions less than 1°, 2°, and 3°. (top) Two sequences with the car image and constant velocity (speeds of 0.75 and 2.5 pixels/frame). (middle) Two sequences with the house image and horizontal speed gradient; first with speeds between 0.65 and 0.85 pixels/frame, and second with speeds between 1.15 and 1.85 pixels/frame. (bottom) Image sequences with a vertical speed gradient, one with the car image and image speeds between 0.875 and 2.125 pixels/frame, and one with the house image, with speeds from 1.07 to 1.93.*

Figure D.3. Component Velocity Error vs. Image Position: *Average component velocity error is shown as a function of image location for the experiments corresponding to Figures D.2 (top-left) and D.2 (middle-left).*

speed gradient introduces motion parallax (curl and shear). In both cases the results are good. The behaviour of errors in terms of distance from filter tuning, estimated orientation and image location is similar to cases above.

Front-View Motion and Image Rotation

The next two experiments involved camera motion along the line of sight. Both cases involved a relatively fast approach toward the surface and a non-zero surface gradient. Figure D.4 shows the error histograms. The error histogram in Figure D.4 *(top-left)* shows the results of a sequence with the car image in which the time to collision is only 48 frames and images speeds are 2.6 at the edges (scene parameters: $\alpha_c = 0$; $\alpha_x = 20$; $\alpha_y = 0$; $d_0 = 12$; $v_c = 0.25$). The error histogram in Figure D.4 *(top-right)* shows the results with the house image (scene parameters: $\alpha_c = 0$; $\alpha_x = 25$; $\alpha_y = 0$; $d_0 = 11.4$; $v_c = 0.2$).

The final two cases involved counter-clockwise image rotation. The rates of rotation were 0.5°/frame for the car image and 1.5°/frame for the house image. The image speeds, with an image size of 150 × 150, were as high as 0.93 and 2.78 pixels/frame respectively. The histograms of errors are given in Figure D.4 *(bottom)*. In both these cases, as in the two above with considerable dilation, a model of image translation is inappropriate near the centre of the image (the fixed point of rotation). In the faster case (Figure D.4 *(bottom-right)*) the variation in speed per pixel as one moves from fixed-point of rotation toward the edge of the images was about 1°/pixel. Over the extent of filter support this means variation of up to 15°. Despite this,

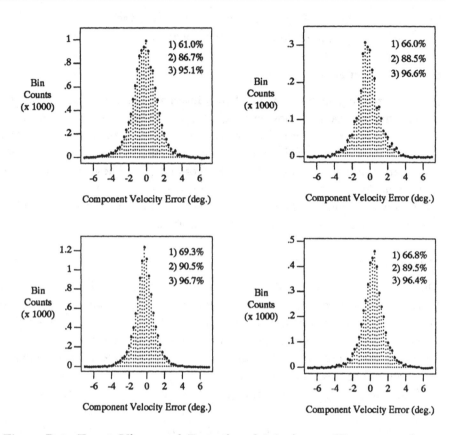

Figure D.4. **Front-View and Rotational Motion:** *Histograms of component velocity errors for front-view motions and rotational motions with the car and house images. (top-left) Image velocity was 0 at the centre of the image, 1.8 pixels/frame on the left, and 2.6 on the right. (top-right) Image velocity was 0 at the centre of the image, 1.4 pixels/frame on the left, and 2.3 pixels/frame on the right. (bottom-left) Rotation of 0.5°/frame with the car image. The induced image speed was up to 0.65 at the edges (0.93 in the corners). (bottom-right) Rotation of 1.5°/frame with the house image. The induced image speed was up to 1.96 at the edges (2.78 in the corners).*

the results are quite respectable.

D.2   2-D Velocity Measurement

Here we show results of the 2-d velocity estimation technique described in
Chapters 7 and 8 applied to real image sequences.

SRI Tree Sequence

The SRI tree sequence (courtesy of SRI International via Sarnoff Research
Center) has the camera moving roughly perpendicular to its line of sight,
looking at clusters of trees. Figure D.5 *(top)* shows three frames from the
sequence. This is a relatively difficult image sequence to extract velocity measurements from because of the sharp changes in depth. These cause dramatic
changes in the local motion field, and problems due to the nonlinearities associated with occlusion. The effects near occlusion boundaries are evident in
the computed 2-d image velocity (shown in Figure D.5).

NASA Coke Sequence

The NASA image sequence (courtesy of NASA-Ames via Sarnoff Research
Center) was created by a camera moving toward the Coke can in Figure
D.6 *(top)*. The motion field is predominantly dilational, and it is reasonably
sparse because of the larger regions with uniform intensity (except for the
sweater in the upper right). The sequence used was a 300 × 300 pixel window
taken out of the original sequence.

The computed 2-d image velocities are depicted in Figure D.6 *(bottom)*.
Note that the component velocities computed from the phase information
are much denser than the 2-d flow. For example, a dense set of component
velocities exists along the pencils, and through the sweater where the texture
is essentially one dimensional.

Rotating Cube Sequence

The rotating Rubik's cube image sequence (courtesy of Richard Szeliski of
DEC, Cambridge Research Lab) contains a cube rotating counter-clockwise
on a turntable, in front of a stationary camera. Three frames from the
sequence are shown in Figure D.7 *(top)*. The motion field associated with the
cube is clear from the estimated 2-d velocities shown in Figure D.7 *(bottom)*.
The velocity estimates are somewhat sparse due to the limited support of the
measurement process and the occurrence of the aperture problem along the
lines and edges of the cube. Image speeds were as large as 1.5 pixels/frame.

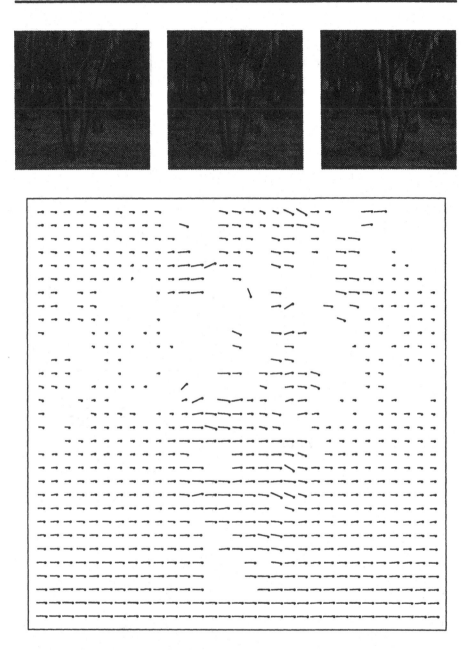

Figure D.5. **2-D Velocity Measurements from SRI Tree Sequence:**
The results of a phase-gradient technique for measuring 2-d image velocity.
(top) Frames 13, 18, and 23 of the tree sequence. (bottom) 2-D velocity
estimates (every sixth pixel) at frame 18.

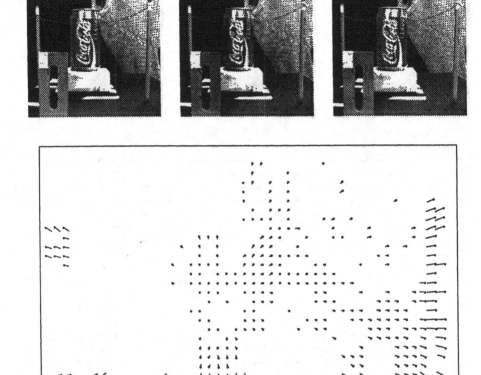

Figure D.6. **2-D Velocity Measurements from NASA Sequence:** *The results of a phase-gradient technique for measuring 2-d image velocity. (top) Frames 7, 12, and 17 of the Coke can sequence. (bottom) 2-D velocity estimates (every sixth pixel) at frame 12.*

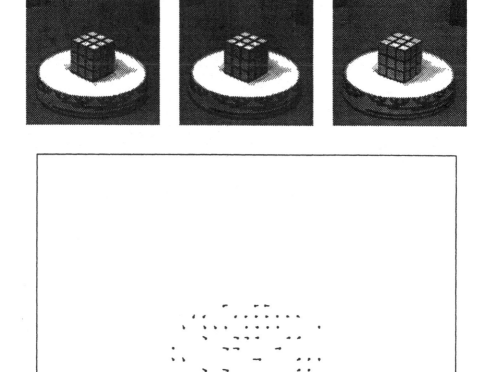

Figure D.7. 2-D Velocity Measurements from Rotating Cube Sequence: *The results of a phase-gradient technique for measuring 2-d image velocity. (top) Frames 10, 15, and 20 of the Rubik's cube image sequence. (bottom) 2-D velocity estimates (every sixth pixel) at frame 15.*

Appendix E

Approximations to $E[\Delta\phi]$ and $E\big[\,|\Delta\phi - E[\Delta\phi]\,|\,\big]$

Let kernels $K_j(x)$ and $H_1(x)$, their respective responses $S_j(x)$ and R_1, and $z_1 = \,<K_0(x), K_1(x)>$ be defined as in Section 9.2. Furthermore, we assume that the real and imaginary parts of the filters $K_j(x)$ are Hilbert transforms of one another. Finally, remember that

$$S_1 \;=\; z_1\, S_0 \,+\, R_1 \,. \tag{E.1}$$

For mean-zero Gaussian white-noise input, the response S_0 is mean-zero Gaussian with variance $\sigma_0^2 \;=\; \| K_0 \|^2 \;=\; 1$ (given stationary white noise the spectral density of the output equals the power spectrum of the filter), and assuming that K_0 is a quadrature filter, the real and imaginary parts of S_0 have a joint Gaussian density, with zero mean and an isotropic covariance $\sigma^2 = 1/2$ [Papoulis, 1965]. The phase of S_0 is uniformly distributed over $(-\pi, \pi]$ and independent of the amplitude. Based on similar arguments, $z_1 S_0$ is also mean-zero Gaussian, but its variance is $|z_1|^2$, and that of its real and imaginary parts is $|z_1|^2/2$ and isotropic. In polar coordinates, $\rho_0 = |z_1 S_0|$ has a Rayleigh density with mean $\mathrm{E}[\rho_0] = \sqrt{\pi}\,|z_1|/2$, and second moment $\mathrm{E}[\rho_0^2] = |z_1|^2$, where $\mathrm{E}[\cdot]$ denotes mathematical expectation. Its phase angle $\psi_0 = \arg[z_1 S_0]$ is independent of ρ_0 and has a uniform density function [Papoulis, 1965].

The real and imaginary parts of the residual process R_1 are Gaussian. But their joint density is not isotropic with uniform phase because the real and imaginary parts of the kernel $H_1(x)$ are not orthogonal; i.e., using the fact that $<\mathrm{Re}[z_1 K_0], \mathrm{Im}[z_1 K_0]>$ and $<\mathrm{Re}[K_1], \mathrm{Im}[K_1]>$ are both 0, it is straightforward to show that

$$<\mathrm{Re}[H_1(x)], \mathrm{Im}[H_1(x)]> \;=\; -\mathrm{Im}[\,<K_1^*(x),\, z_1 K_0(x)>\,]\,, \tag{E.2}$$

which is generally non-zero. As an approximation we assume that $H_1(x)$ is a quadrature filter so that its output is mean-zero with variance $\sigma_1^2 =$

$\| H_1(x) \|^2$. In other words, we assume that $\rho_1 = |R_1|$ has a Rayleigh density with mean $E[\rho_1] = \sqrt{\pi} \| H_1(x) \|/2$, and second moment $E[\rho_1^2] = \| H_1(x) \|^2$. Its phase angle $\psi_1 = \arg[R_1]$ is independent of ρ_1 and has a uniform density function over $(-\pi, \pi]$. These expressions can be simplified further because $\| H_1(x) \|$ can be shown to reduce to $\sqrt{1 - |z_1|^2}$. This is easily derived from $H_1(x) = K_1(x) - z_1 K_0(x)$ as follows:

$$
\begin{aligned}
\| H_1(x) \|^2 &= \ <K_1(x) - z_1 K_0(x),\ K_1(x) - z_1 K_0(x)> \\
&= \ \| K_1(x) \|^2 + \| z_1 K_0(x) \|^2 - \\
&\quad (<z_1 K_0(x),\ K_1(x)> + <K_1(x),\ z_1 K_0(x)>) \\
&= \ 1 + |z_1|^2 - (<z_1 K_0(x),\ K_1(x)> + <z_1 K_0(x),\ K_1(x)>^*) \\
&= \ 1 + |z_1|^2 - 2\,\mathrm{Re}[z_1^* <K_0(x),\ K_1(x)>] \\
&= \ 1 - |z_1|^2 .
\end{aligned}
\tag{E.3}
$$

Finally, because the kernels $z_1 K_0(x)$ and $H_1(x)$ are orthogonal (by construction) the two signals $z_1 S_0$ and R_1 are uncorrelated; and because the input is Gaussian they are independent. With the assumption that R_1 is isotropic in its real and imaginary parts, it has no influence on the mean phase difference $\Delta\phi = \arg[S_1] - \arg[S_0]$. Thus, we approximate $E[\Delta\phi]$ by $\tilde{\mu}(z_1) = \arg[z_1]$. By the same argument, the variation of $\Delta\phi$ about the mean is determined by the phase difference between $z_1 S_0$ and $S_1 = z_1 S_0 + R_1$; that is

$$
\begin{aligned}
\Delta\phi - \tilde{\mu}(z_1) &= \ \arg[S_1] - \arg[S_0] - \arg[z_1] \\
&= \ \arg[z_1 S_0 + R_1] - \arg[z_1 S_0] .
\end{aligned}
\tag{E.4}
$$

When $|R_1| < |z_1 S_0|$ (i.e. close to $\mathbf{p}_0$), the magnitude of $\Delta\phi - \tilde{\mu}(z_1)$ is given by the magnitude of the arctangent of the component of R_1 that is perpendicular to the complex direction of $z_1 S_0$, divided by the magnitude of $z_1 S_0$ (cf. Figure 9.2); that is,

$$
|\tilde{\Delta\phi}| = \left| \arctan\left(\frac{d_1}{\rho_0} \right) \right| \le \frac{|d_1|}{\rho_0} ,
\tag{E.5}
$$

where $\rho_0 = |z_1 S_0|$, and

$$
d_1 = \frac{\mathrm{Im}[(z_1 S_0)^* R_1]}{|z_1 S_0|} = \mathrm{Im}[e^{-i\psi_0} R_1] = \rho_1\, \mathrm{Im}[e^{i(\psi_1 - \psi_0)}] ,
\tag{E.6}
$$

where $\psi_0 = \arg[z_1 S_0]$ and $R_1 = \rho_1 e^{i\psi_1}$. With complex numbers represented as vectors in the complex plane, d_1 amounts to the length of the projection of R_1 onto the unit vector normal to $z_1 S_0$. Finally, we can now formulate a bound on $E[|\tilde{\Delta\phi}|]$

$$
E[|\tilde{\Delta\phi}|] \le E\left[\frac{|d_1|}{\rho_0} \right] = E[\rho_0^{-1}]\, E[\rho_1]\, E[|\sin(\psi_1 - \psi_0)|] .
\tag{E.7}
$$

This follows from the independence of the four random quantities ρ_0, ρ_1, ψ_0, and ψ_1. Given the Rayleigh density functions for ρ_0 and ρ_1, and uniform density functions for ψ_0 and ψ_1, it can be shown that $E[\rho_0^{-1}] = \sqrt{\pi}/|z_1|$ and $E[|\sin(\psi_1 - \psi_0)|] = 2/\pi$. Therefore, we can rewrite the bound as

$$E[|\tilde{\Delta}\phi|] \leq \frac{\sqrt{1 - |z_1|^2}}{|z_1|} . \qquad (E.8)$$

It is tightest for small values of $\tilde{\Delta}\phi$ because of the bound used in (E.5).

Appendix F

Derivations of z_1

Chapter 9 outlines the decomposition of a kernel K_1 into two components: one in the span of K_0, and the other orthogonal to K_0. The essential ingredient of this decomposition is given by z_1, the inner product of the generic kernel K_0 with deformations of itself given by K_1. This appendix derives expressions for z_1 for the cases of Gabor kernels (9.8) and modulated square-wave kernels (9.23).

F.1   Gabor Kernels

In the first case, let

$$z_1 \;=\; <G_0(x),\, G_1(x)> \;, \qquad\qquad (F.1)$$

where $G_j(x) \;=\; Gabor(x - x_j; \sigma_j, k_j)$ denotes a Gabor kernel (9.8) at location $\mathbf{p}_j = (x_j, \lambda_j)^T$. Here, $\sigma_j = \sigma(\lambda_j)$ and $k_j = k(\lambda_j)$ as in (9.10). Let $\Delta x = x_1 - x_0$, and let $\Delta k = k_1 - k_0$. Without loss of generality let $x_0 = 0$.

In the case of Gabor kernels with a constant octave bandwidth, we can use Parseval's theorem to solve for z_1 as follows:

$$
\begin{aligned}
<G_0(x),\, G_1(x)> \;&=\; (2\pi)^{-1} <\hat{G}_0(k),\, \hat{G}_1(k)> \\
&=\; (2\pi)^{-1} \int \hat{G}_0(k)^* \, \hat{G}_1(k)\, dk \\[2mm]
&=\; \frac{\sqrt{\pi \sigma_0 \sigma_1}}{\pi} \int e^{-(k-k_0)^2\, \sigma_0^2/2}\; e^{-(k-k_1)^2\, \sigma_1^2/2}\; e^{i\Delta x\, k}\, dk\,. \quad (F.2)
\end{aligned}
$$

With a change of variables $(\omega = k - k_0)$, and straightforward algebraic manipulation (complete the square) this becomes

$$\frac{\sqrt{\pi \sigma_0 \sigma_1}}{\pi}\, e^{i\Delta x\, k_0}\, e^{-\Delta k^2\, \sigma_1^2\,[1-(\sigma_1^2/\bar{\sigma}^2)]/2} \int e^{-\bar{\sigma}^2\,[\omega - (\Delta k\, \sigma_1^2/\bar{\sigma}^2)]^2/2}\; e^{i\Delta x\, \omega}\, d\omega\,, \quad (F.3)$$

where $\bar{\sigma}^2 = \sigma_0^2 + \sigma_1^2$. In (F.3) the integrand is a Gabor function and therefore easily integrated. After solving the integral, and a small amount of algebraic manipulation (using $\sigma_1^2 \bar{k}^2 = \bar{\sigma}^2 k_0^2$, where $\bar{k}^2 = k_0^2 + k_1^2$), one can show that

$$<G_0(x), G_1(x)> \quad = \quad \frac{\sqrt{2\sigma_0\sigma_1}}{\bar{\sigma}} e^{-\Delta k^2 \sigma_0^2 \sigma_1^2 / 2\bar{\sigma}^2} e^{-\Delta x^2 / 2\bar{\sigma}^2} e^{i\Delta x [k_0 + (\Delta k\, k_0^2 / \bar{k}^2)]}$$

$$= \quad \sqrt{2\pi}\, G(\Delta x; \bar{\sigma})\, G(\Delta k; \bar{\sigma}/\sigma_0\sigma_1)\, e^{i\,\Delta x\, [k_0 + (\Delta k\, k_0^2 / \bar{k}^2)]}$$

where $G(x; \sigma)$ is the scaled Gaussian in (9.9).

In the case of Gabor kernels for which the support width σ is constant (i.e. independent of λ so that the *linear* bandwidth is constant), z_1 is given by

$$z_1 \quad = \quad \frac{4\pi}{\sigma} e^{i\,\Delta x\, [k_0 + (\Delta k/2)]} e^{-\Delta k^2 \sigma^2 / 4} e^{-\Delta x / 4\sigma^2}\,.$$

F.2   Modulated Square-Wave Filter

The modulated square-wave kernel is given by

$$K(x, \lambda) \quad = \quad e^{ik(\lambda)x}\, sq(x, \lambda)\,, \tag{F.4}$$

where $k(\lambda) = 2\pi/\lambda$ is the peak tuning frequency, and

$$sq(x, \lambda) \quad = \quad \begin{cases} \frac{1}{\sqrt{\lambda}} & \text{if } |x| \leq \lambda/2 \\ 0 & \text{otherwise}\,. \end{cases} \tag{F.5}$$

Then, as above, with $\lambda_1 = \lambda_0 + \Delta\lambda$, $k_j = k(\lambda_j)$, and $\Delta k = k_1 - k_0$,

$$z_1 \quad = \quad \int_{-\infty}^{\infty} e^{-ik_0 x}\, e^{i(k_0 + \Delta k)(x + \Delta x)}\, sq(x, \lambda)\, sq(x + \Delta x, \lambda_1)\, dx$$

$$= \quad \frac{1}{\sqrt{\lambda_0\lambda_1}} e^{i\Delta x k_1} \int_a^b e^{i\Delta k x}\, dx$$

$$= \quad i\, e^{i\Delta x k_1} \left(e^{i\Delta k a} - e^{i\Delta k b} \right) \left(\Delta k \sqrt{\lambda_0\lambda_1} \right)^{-1}\,, \tag{F.6}$$

where

$$b \quad = \quad \frac{\lambda_0}{2} + \min\left[0,\, \frac{\Delta\lambda}{2} - \Delta x \right]\,, \qquad a \quad = \quad \frac{-\lambda_0}{2} + \max\left[0,\, -\frac{\Delta\lambda}{2} - \Delta x \right]\,.$$

In the limit as $\Delta k \to 0$, z_1 converges to $e^{i\Delta x k_0}\, (b - a)\, \lambda_0^{-1}$.

Appendix G

Density Functions for $\phi_x(x)$ and $\rho_x(x)/\rho(x)$

Given a band-pass complex-valued signal $S(x)$ and its derivative $S_x(x)$, the local (instantaneous) frequency of S, defined by $\tilde{k}(x) \equiv \phi_x(x)$ [Papoulis, 1965], is given by

$$\tilde{k}(x) = \frac{\text{Im}[S(x)^* S_x(x)]}{|S(x)|^2}$$

$$= \frac{\text{Re}[S(x)]\,\text{Im}[S_x(x)] - \text{Im}[S(x)]\,\text{Re}[S_x(x)]}{\text{Im}[S(x)]^2 + \text{Re}[S(x)]^2}. \tag{G.1}$$

Similarly, the relative amplitude derivative, $\tilde{\rho}(x) \equiv \rho_x(x)/\rho(x)$, can be expressed as

$$\tilde{\rho}(x) = \frac{\text{Re}[S(x)^* S_x(x)]}{|S(x)|^2}$$

$$= \frac{\text{Re}[S(x)]\,\text{Re}[S_x(x)] + \text{Im}[S(x)]\,\text{Im}[S_x(x)]}{\text{Im}[S(x)]^2 + \text{Re}[S(x)]^2}. \tag{G.2}$$

In the complex plane, (G.1) corresponds to a scaled projection of $S_x(x)$ onto the unit vector orthogonal to $S(x)$. Similarly, (G.2) corresponds to the scaled projection of $S_x(x)$ onto $S(x)$.

Let $S(x)$ be the response of a band-pass, differentiable kernel $K(x, \lambda)$ to mean-zero, white Gaussian noise input denoted $I(x)$. Thus $S(x)$ may be expressed as

$$S(x) = \ <K^*(x), I(x)> . \tag{G.3}$$

Differentiating (G.3) we find that $S_x(x)$ is given by

$$S_x(x) = \ <K_x^*(x), I(x)> . \tag{G.4}$$

With the assumption of Gaussian noise, the real and imaginary parts of $S(x) = a_1 + ib_1$ and $S_x(x) = a_2 + ib_2$ have a 4-d joint Gaussian distribution. The probability density function is

$$p(\mathbf{s}) = \frac{1}{4\pi^2 |C|^{1/2}} e^{-0.5\mathbf{s}^t C^{-1}\mathbf{s}} , \tag{G.5}$$

where $\mathbf{s} = (a_1, b_1, a_2, b_2)^T$, C denotes the covariance matrix, and $|C|$ denotes its determinant. The covariance matrix has the form:

$$C = \begin{pmatrix} s_1^2 & 0 & c_1 & c_2 \\ 0 & s_1^2 & -c_2 & c_1 \\ c_1 & -c_2 & s_2^2 & 0 \\ c_2 & c_1 & 0 & s_2^2 \end{pmatrix} , \tag{G.6}$$

with determinant $|C| = (s_1^2 s_2^2 - c_1^2 - c_2^2)^2$. The individual elements of C are given by

$$s_1 = \| \operatorname{Re}[K(x)] \| ,$$
$$s_2 = \| \operatorname{Re}[K_x(x)] \| ,$$

$$c_1 = \; < \operatorname{Re}[K(x)], \operatorname{Re}[K_x(x)] > \; ,$$
$$c_2 = \; < \operatorname{Re}[K(x)], \operatorname{Im}[K_x(x)] >$$

The form of the matrix (e.g. its zeros and symmetries) can be shown easily using Parseval's Theorem, and the fact that the real and imaginary parts of each kernel form quadrature pairs (i.e. they are Hilbert transforms of one another).

For the derivation of the probability density functions for $\tilde{k}$ and $\tilde{\rho}$ it is convenient to adopt polar coordinates with the following change of variables:

$$(a_1, b_1) = \frac{|C|^{1/4} r_1}{s_2} (\cos\theta_1, \sin\theta_1) ,$$

$$(a_2, b_2) = \frac{|C|^{1/4} r_2}{s_1} (\cos\theta_2, \sin\theta_2) .$$

The determinant of the Jacobian is $|C| r_1 r_2 / s_1^2 s_2^2$. In these coordinates, the density function simplifies to

$$pdf(r_1, \theta_1, r_2, \theta_2) = A \exp\left(-\frac{1}{2}\left[r_1^2 + r_2^2 - \frac{2 r_1 r_2}{s_1 s_2}(c_2 \sin\Delta\theta + c_1 \cos\Delta\theta)\right]\right),$$

where $\Delta\theta = \theta_2 - \theta_1$ and $A = |C|^{1/2} r_1 r_2 / 4\pi^2 s_1^2 s_2^2$. Furthermore, the local frequency (G.1) and relative amplitude derivative (G.2) become

$$\tilde{k} = \frac{s_2 r_2}{s_1 r_1} \sin\Delta\theta , \qquad \tilde{\rho} = \frac{s_2 r_2}{s_1 r_1} \cos\Delta\theta .$$

In solving for the density functions for $\tilde{k}$ and $\tilde{\rho}$, following [Broman, 1981], we first determine their probability distribution functions. For example, the distribution function for $\tilde{k} < 0$ can be written as

$$P(\tilde{k}) = \int_0^{2\pi} d\theta_2 \int_{\theta_2}^{\theta_2+\pi} d\theta_1 \int_0^{\infty} dr_1 \int_B^{\infty} dr_2 \; pdf(r_1, \theta_1, r_2, \theta_2) , \quad (G.7)$$

where $B = r_1 s_1 \tilde{k}/s_2 \sin \Delta\theta$, and the limits of integration are derived from $(s_2 r_2 / s_1 r_1) \sin \Delta\theta < \tilde{k}$ and $\tilde{k} < 0$. Upon differentiating (G.7) with respect to $\tilde{k}$, and with some manipulation of integrals, the density function for $\tilde{k}$ can be reduced to

$$pdf(\tilde{k}) = \frac{|\tilde{k}| |C|^{1/2}}{\pi} \int_0^{-\pi} \frac{\sin^2 \psi \; d\psi}{(s_1^2 \tilde{k}^2 + (s_2^2 - 2c_2\tilde{k}) \sin^2 \psi - c_1\tilde{k} \sin 2\psi)^2} .$$

Similarly, the probability density function for $\tilde{\rho}$ can be shown to be

$$pdf(\tilde{\rho}) = \frac{|\tilde{\rho}| |C|^{1/2}}{\pi} \int_{\pi/2}^{-\pi/2} \frac{\cos^2 \psi \; d\psi}{(s_1^2 \tilde{\rho}^2 + (s_2^2 - 2c_1\tilde{\rho}) \cos^2 \psi - c_2\tilde{\rho} \sin 2\psi)^2} .$$

Bibliography

Adelson, E.H. and Bergen, J.R. (1985) Spatiotemporal energy models for the perception of motion. *J. Opt. Soc. Am. A* 2, pp. 284-299

Adelson, E.H. and Bergen, J.R. (1986) The extraction of spatiotemporal energy in human and machine vision. *Proc. IEEE Workshop on Motion*, Charleston, pp. 151-156

Adelson, E.H. and Movshon, J.A. (1982) Phenomenal coherence of moving visual patterns. *Nature* 300, pp. 523-525

Adelson, E.H., Simoncelli, E., and Hingorani, R. (1987) Orthogonal pyramid transforms for image coding. *Proc. SPIE Conf. Vis. Comm. Im. Proc.*, Cambridge, pp. 50-58

Adiv, G. (1985) Determining three-dimensional motion and structure from optical flow generated by several moving objects. *IEEE Trans. PAMI* 7, pp. 384-401

Anandan, P. (1989) A computational framework and an algorithm for the measurement of visual motion. *Int. J. Computer Vision* 2, pp. 283-310

Barron, J.L., Jepson, A.D., and Tsotsos, J.K. (1990) The Feasibility of Motion and Structure from Noisy Time-Varying Image Velocity Information. *Int. J. Computer Vision* 5, pp. 239-269

Bergholm, F. (1988) A theory of optical velocity fields and ambiguous motion of curves. *Proc. IEEE ICCV*, Tampa, pp. 165-176

Bertero, M., Poggio, T.A., and Torre, V. (1988) Ill-posed problems in early vision. *Proc. IEEE* 76, pp. 869-889

Bracewell, R.N. (1978) *The Fourier Transform and its Applications.* McGraw-Hill, New York

Broman, H. (1981) The instantaneous frequency of a Gaussian signal: The one-dimensional density function. *IEEE Trans. ASSP* 29, p. 108-111

Burt, P.J. (1981) Fast filter transforms for image processing. *Computer Graphics and Image Processing* 16, pp. 20-51

Burt, P.J. and Adelson, E.H. (1983) The Laplacian pyramid as a compact image code. *IEEE Trans. Commun.* 31, pp. 532-540

Burt, P.J., Yen, C., and Xu, X. (1983) Multiresolution flow-through motion analysis. *Proc. IEEE CVPR*, Washington, pp. 246-252

Burt, P.J., Bergen, J., Hingorani, R., Kolczynski, R., Lee, W., Leung, A., Lubin, J., and Shvaytser, H. (1989) Object tracking with a moving camera. *Proc. IEEE Workshop on Visual Motion*, Irvine, pp. 2-12.

Buxton, B.F. and Buxton, H. (1984) Computation of optic flow from the motion of edge features in image sequences. Image and Vision Computing 2, pp. 59-75

Canny, J.F. (1986) A computational approach to edge detection. *IEEE Trans. PAMI* 8, pp. 679-698

Cook, R.L. and Torrance, K.E. (1982) A reflectance model for computer graphics. *ACM Trans. Graphics* 1, pp. 7-24

Crowley, J.L. and Parker, A.C. (1984) A representation for shape based on peaks and troughs in the difference of low-pass transform. *IEEE Trans. PAMI* 6, pp. 156-169

Daugman, J.G. (1987) Pattern and motion vision without Laplacian zero crossings. *J. Opt. Soc. Am. A* 5, pp. 1142-1148.

Davenport, W.B. and Root, W. (1958) *Introduction to the Theory of Random Signals and Noise*, McGraw-Hill, New York

Dudgeon, D.E. and Mersereau, R.M. (1984) *Multidimensional Digital Signal Processing*, Prentice-Hall, New Jersey

Duncan, J.H. and Chou, T.C. (1988) Temporal edges: the detection of motion and the computation of optical flow. *Proc. 2nd IEEE ICCV*, Tampa, pp. 374-382

Dutta, R., Manmatha, R., Williams, L., and Riseman, E.M. (1989) A data set for quantitative motion analysis. *Proc. IEEE CVPR*, San Diego, pp. 159-164

Enkelmann, W. (1986) Investigations of multigrid algorithms for the estimation of optical flow fields in image sequences. *Proc. IEEE Workshop on Motion*, Charleston, pp. 81-87

Fahle, M., and Poggio, T. (1981) Visual hyperacuity: spatiotemporal interpolation in human vision. *Proc. R. Soc. Lond.* B213, pp. 451-477.

Faugeras, O. (1990) On the motion of 3-d curves and its relationship to optical flow. *Proc. 1st ECCV*, Antibes, Springer-Verlag, pp. 107-117

Fennema, C. and Thompson, W. (1979) Velocity determination in scenes containing several moving objects. *Computer Graphics and Image Processing* 9, pp. 301-315

Fleet, D.J. (1988) Implementation of velocity-tuned filters and image encoding. Technical Report: FBI-HH-M-159/88, Dept. Computer Science, University of Hamburg

Fleet, D.J. and Jepson, A.D. (1985) Velocity extraction without form interpretation. *Proc. IEEE Workshop on Computer Vision*, Bellaire, pp. 179-185

Fleet, D.J. and Jepson, A.D. (1989) Hierarchical construction of orientation and velocity selective filters. *IEEE Trans. PAMI* 11, pp. 315-325

Fleet, D.J. and Jepson, A.D. (1990) Computation of component image velocity from local phase information. *Int. J. Computer Vision* 5, pp. 77-104

Fleet, D.J. and Jepson, A.D. (1991) Stability of Phase Information. *Proc. IEEE Workshop on Visual Motion*, Princeton, pp. 52-60

Fleet, D.J., Jepson, A.D. and Jenkin, M. (1991) Phase-based disparity measurement. *CVGIP: Image Understanding* 53, pp. 198-210

Gabor, D. (1946) Theory of communication. *J. IEE* 93, pp. 429-457

Gallager, R.G. (1968) *Information Theory and Reliable Communication.* J. Wiley and Sons, New York

Gardenhire, L.W. (1964) Selecting sampling rates. *Proc. 19th Instrument Soc. Am. Conf.*, July 1964

Girod, B. and Kuo, D. (1989) Direct estimation of displacement histograms. *Proc. OSA Topcial Meeting on Image Understanding and Machine Vision*, pp. 73-76, Cape Cod

Girosi, F., Verri, A., and Torre, V. (1989) Constraints for the computation of optical flow. *Proc. IEEE Workshop on Visual Motion*, Irvine, pp. 116-124

Glazer, F. (1981) Computing optic flow. *Proc. IJCAI*, Vancouver, pp. 644-647

Glazer, F. (1987) Hierarchical gradient-based motion detection. *Proc. DARPA Image Understanding Workshop*, Los Angeles, California, pp. 733-748

Glazer, F., Reynolds, G., and Anandan, P. (1983) Scene matching through hierarchical correlation. *Proc. IEEE CVPR*, Washington, pp. 432-441

Gong, S. (1989) Curve motion constraint equation and its applications. *Proc. IEEE Workshop on Visual Motion*, Irvine, pp. 73-80

Grzywacz, N.M. and Yuille, A.L. (1990) A model for the estimate of local image velocity by cells in the visual cortex. *Proc. R. Soc. Lond.* B 239, pp. 129-161

Heeger, D.J. (1987) A model for the extraction of image flow. *J. Opt. Soc. Am. A* 4, pp. 1455-1471

Heeger, D.J. (1988) Optical flow using spatiotemporal filters. *Int. J. Computer Vision* 1, pp. 279-302

Heeger, D.J. and Jepson, A.D. (1990) Simple method for computing 3d motion and depth. *Proc. 3rd IEEE ICCV*, Osaka, pp. 96-100

Hildreth, E.C. (1984) The computation of the velocity field. *Proc. R. Soc. Lond.* B 221, pp. 189-220

Horn, B.K.P. (1977) Understanding image intensities. *Artificial Intelligence* 8, pp. 201-231

Horn, B.K.P. (1986) *Robot Vision*. MIT Press, Cambridge

Horn, B.K.P. and Schunck, B.G. (1981) Determining optic flow. *Artificial Intelligence* 17, pp. 185-204

Horn, B.K.P. and Sjoberg, R.W. (1979) Calculating the reflectance map. *Applied Optics* 18, pp. 1770-1779

Hou, H.S. and Andrews, H.C. (1978) Cubic splines for image interpolation and digital filtering. *IEEE Trans. ASSP* 26, pp. 508-517

Huber, P.J. (1981) *Robust Statistics*, John Wiley & Sons, New York

Jahne, B. (1990) Motion determination in space-time images. *Proc. 1st ECCV*, Antibes, Springer-Verlag, pp. 161-173

Jenkin, M., and Jepson, A.D. (1988) The measurement of binocular disparity. In *Computational Processes in Human Vision*, edited by Z. Pylyshyn, Ablex Press, New Jersey

Jepson, A.D., and Fleet, D.J. (1990) Scale-space singularities. *Proc. 1st ECCV*, Antibes, Springer-Verlag, pp. 50-55

Jepson, A.D. and Heeger, D.J. (1990) Subspace methods for recovering rigid motion, Part II: Theory. Technical Report: RBCV-TR-90-36, Department of Computer Science, University of Toronto

Jepson, A.D. and Jenkin, M. (1989) Fast computation of disparity from phase differences. *Proc. IEEE CVPR*, San Diego, pp. 398-403

Jepson, A.D. and Richards, W. (1990) What is a Percept? (manuscript in preparation)

Julesz, B. (1971) *Foundations of Cyclopean Perception*, Univ. of Chicago Press, Chicago

Kass, M., Witkin, A., and Terzopoulos, D. (1988) Snakes: Active contour models. *Int. J. Computer Vision* 1, pp. 321-331

Kearney, J.K., Thompson, W.B., and Boley, D.L. (1987) Optical flow estimation: An error analysis of gradient-based methods with local optimization. *IEEE Trans. PAMI* 9, pp. 229-244

Kendall, M.G. (1961) *The Geometry of n Dimensions*, C. Griffin Co., London.

Koenderink, J.J. (1984) The structure of images. *Biological Cybernetics* 50, pp. 363-370

Koenderink, J.J. and van Doorn, A.J. (1976) Local structure of movement parallax of the plane. *J. Opt. Soc. Am.* 66, pp. 717-723

Koenderink, J.J. and van Doorn, A.J. (1987) Facts on optic flow. *Biological Cybernetics* 56, pp. 247-254

Kolers, P.A. (1972) *Aspects of Motion Perception*. Pergamon Press, Oxford

Kuglin, C. and Hines, D. (1975) The phase correlation image alignment method. *Proc. IEEE Int. Conf. Cybern. Society*, pp. 163-165

Langer, I.M.S. (1988) On efficient representation of natural images. M.Sc. Thesis, Dept. of Computer Science, University of Toronto

Langley, K., Atherton, T., Wilson, R., and Larcombe, M. (1990) Vertical and horizontal disparities from phase. *Proc. 1st ECCV*, Antibes, Springer-Verlag, pp. 315-325

Lindeberg, T. (1990) Scale-space for discrete signals. *IEEE Trans. PAMI* 12, pp. 234-254

Little, J.J., Bulthoff, H.H., and Poggio, T.A. (1988) Parallel optical flow using local voting. *Proc. 2nd IEEE ICCV*, Tampa, pp. 454-459

Little, J.J. and Verri, A. (1989) Analysis of differential and matching methods for optical flow. *IEEE Workshop on Visual Motion*, Irvine, pp. 173-180

Longuet-Higgins, H.C., and Prazdny, K. (1980) The interpretation of a moving retinal image. *Proc. R. Soc. Lond.* B 208, pp. 385-397

Lowe, D. (1985) *Perceptual Organization and Visual Recognition.* Kluwer Academic Publishers, Hingham, MA

Mallat, S.G. (1989) Multifrequency channel decomposition of images and wavelet models. *IEEE Trans. ASSP* 37, pp. 2091-2110

Marr, D. (1982) Vision. W.H. Freeman and Co., New York

Marr, D. and Hildreth, E., (1980) Theory of edge detection. *Proc. R. Soc. Lond.* B 207, pp. 187-217

Marr, D. and Poggio, T. (1979) A computational theory of human stereo vision. *Proc. R. Soc. Lond.* B204, pp. 301-328

Marr, D. and Ullman, S. (1981) Directional selectivity and its use in early visual processing. *Proc. R. Soc. Lond.* B 211, pp. 151-180

Mayhew, J. and Frisby, J. (1981) Computational studies toward a theory of human stereopsis. *Artificial Intelligence* 17, pp 349-385

Morgan, M.J. (1980) Analogue models of motion perception. *Phil. Trans. R. Soc. (Lond.)* B290, pp. 117-135

Morrone, M.C. and Burr, D.C. (1988)   Feature detection in human vision: a phase-dependent energy model. *Proc. R. Soc. Lond.* B 235, pp. 221-245

Movshon, J.A., Adelson, E., Gizzi, M., and Newsome, W. (1985) The analysis of moving visual patterns. in *Pattern Recognition Mechanisms* (eds.) Chages, C., Gattass, R, and Gross, C. Vatican Press, Rome, pp. 117-151

Nagel, H.H. (1983) Displacement vectors derived from second-order intensity variations in image sequences. *Computer Graphics and Image Processing* 21, pp. 85-117

Nagel, H.H. (1987) On the estimation of optical flow: Relations between different approaches and some new results. *Artificial Intelligence* 33, pp. 299-324

Nagel, H.H. (1989) On a constraint equation for the estimation of displacement rates in image sequences. *IEEE Trans. PAMI* 11, pp. 13-30

Nagel, H.H. and Enkelmann, W. (1986) An investigation of smoothness constraints for the estimation of displacement vector fields from image sequences. *IEEE Trans. PAMI* 8, pp 565-593

Nakayama, K. and Silverman, G.H. (1988a) The aperture problem – I: Perception of nonrigidity and motion direction in translating sinusoidal lines. *Vision Research* 28, pp. 739-746

Nakayama, K. and Silverman, G.H. (1988b) The aperture problem – II: Spatial integration of velocity information along contours. *Vision Research* 28, pp. 747-753

Netravali, A.N. and Limb, J.O. (1980) Picture coding: A review. *Proc. IEEE* 68, pp. 366-406

Nishihara, K. (1984) Prism: A practical real-time imaging stereo matcher. A.I. Memo 780, AI Lab, MIT

Ogle, K.N. (1956) *Research in Binocular Vision*, W.B. Saunders Co., Philadelphia

Olson, T. and Potter, R. (1989) Real-time vergence control. *Proc. IEEE CVPR*, San Diego, pp. 404-409

Oppenheim, A.V. and Schafer, R.W. (1975) *Digital Signal Processing.* Prentice-Hall, Englewood Cliffs

Papoulis, A. (1965) *Probability, Random Variables, and Stochastic Processes*, McGraw-Hill, New York

Papoulis, A. (1968) *Systems and Transforms with Applications in Optics*, McGraw-Hill, New York

Pentland, A.P. (1986) Perceptual organization and the representation of natural form. *Artificial Intelligence* 28, pp. 293-331

Phong, B.T. (1975) Illumination for computer generated pictures. *Comm. ACM* 18, pp. 311-317

Priestley, H.A. (1985) *Introduction to Complex Analysis*, Clarendon Press, Oxford

Ramachandran, V.S. and Anstis, S.M. (1983) Displacement thresholds for coherent apparent motion random dot-patterns. *Vision Research* 24, pp. 1719-1724

Ramachandran, V.S. and Cavanagh, P. (1987) Motion capture anisotropies. *Vision Research* 27, pp. 97-106

Sanger, T. (1988) Stereo disparity computation using Gabor filters. *Biological Cybernetics* 59, pp. 405-418

Santen, J.P.H. van and Sperling, G. (1985) Elaborated Reichardt detectors. *J. Opt. Soc. Am A* 2, pp. 300-321

Schafer, R.W. and Rabiner, L.R. (1973) A digital signal approach to interpolation. *Proc. IEEE* 61, pp. 692-702

Shannon, C.E. and Weaver, W. (1963) *The Mathematical Theory of Communication.* University of Illinois Press, Urbana

Shizawa, M. and Mase, K. (1990) Simultaneous multiple optical flow estimation. *Proc. IEEE ICPR*, Atlantic City, pp. 274-278

Slepian, D. (1976) On bandwidth. *Proc. IEEE* 64, pp. 292-300

Slepian, D. (1983) Some comments on Fourier analysis, uncertainty and modelling. *Siam Review* 25, pp. 379-393

Tretiak, O. and Pastor, L. (1984) Velocity estimation from image sequences with second order differential operators. *Proc. IEEE ICPR*, Montreal, pp. 20-22

Torre, V. and Poggio, T.A. (1986)   On edge detection. *IEEE Trans. PAMI* 8, pp. 147-163

Ty, K.M., and Venetsanopoulos, A.N. (1984)   Sampling non-band-limited signals. *Proc. Telecon '84*, Halkidiki, Greece.

Uras, S., Girosi, F., Verri, A., and Torre, V. (1989) A computational approach to motion perception. *Biological Cybernetics* 60, pp. 79-87

Verri, A. and Poggio, T. (1987) Against quantitative optical flow. *Proc. IEEE ICCV*, London, pp. 171-180

Watson, A.B. and Ahumada, A.J. (1983) A look at motion in the frequency domain. Technical Report: 84352, NASA-Ames Research Center.

Watson, A.B. and Ahumada, A.J. (1985) Model of human visual-motion sensing. *J. Opt. Soc. Am. A* 2, pp. 322-342

Waxman, A.M. and Ullman, S. (1985)   Surface structure and three-dimensional motion from image flow kinematics. *Int. J. Robotics Res.* 4, pp. 72-94

Waxman, A.M. and Wohn, K. (1985) Contour evolution, neighbourhood deformation and global image flow: Planar surfaces in motion. *Int. J. Robotics Res.* 4, pp. 95-108

Waxman, A.M., Wu, J., and Bergholm, F. (1988) Convected activation profiles: Receptive fields for real-time measurement of short-range visual motion. *Proc. IEEE CVPR*, Ann Arbor, pp. 717-723

Whitted, T. (1980) An improved illumination model for shaded display. *Comm. ACM* 23, pp. 343-349

Williams, D.W. and Sekuler, R. (1984) Coherent global motion percepts from stochastic local motions. *Vision Research* 24, pp. 55-62

Witkin, A.P. (1983) Scale-space filtering. *Proc. 8th IJCAI*, Karlsruhe, pp. 1019-1022

Witkin, A.P. and Tenenbaum, J.M. (1983) On the role of structure in vision. in *Human and Machine Viison*, editted by Beck, J., Hope, B., and Rosenfeld, A., Academic Press

Witkin, A.P., Terzopoulos, D., and Kass, M. (1987) Signal matching through scale-space. *Int. J. of Computer Vision* 1, pp. 133-144

Wu, J., Brockett, R. and Wohn, K. (1989) A contour-based recovery of image flow: Iterative method. *Proc. IEEE CVPR*, San Diego, pp. 124-129

Xu, L., Oja, E., and Kultanen, P. (1990) Randomized Hough transform (RHT): Theoretical analysis and extensions. *Computer Vision, Graphics and Image Processing* (submitted)

Yeshurun, Y. and Schwartz, E. (1989) Cepstral filtering on a columnar image architecture: A fast algorithm for binocular stereo segmentation. *IEEE Trans. PAMI* 11, pp. 759-767

Yuille, A.L. and Grzywacz, N.M. (1988) The motion coherence theory. *Proc. IEEE ICCV*, Tampa, pp. 344-353

Yuille, A.L. and Poggio, T.A. (1986) Scaling theorems for zero-crossings. *IEEE Trans. PAMI* 8, pp. 15-25

Zucker, S.W. (1981) Computer vision and human perception: An essay on the discovery of constraints. *Proc. IJCAI*, Vancouver, pp. 1102-1116

Zucker, S.W. and Iverson, L. (1987) From orientation selection to optical flow. *Computer Vision, Graphics and Image Processing* 37, pp. 196-220

Index